Urban Biodiversity

Exploring Natural Habitat and its Value in Cities

Valentin Schaefer
Hillary Rudd
Jamie Vala

Captus Press

Library and Archives Canada Cataloguing in Publication

Schaefer, Valentin, 1951–
Urban biodiversity : exploring natural habitat and its value in cities / by Valentin Schaefer, Hillary Rudd and Jamie Vala.

Includes bibliographical references and index.
ISBN 1-55322-078-1

1. Biological diversity conservation. 2. Conservation of natural resources. 3. Urbanization. I. Rudd, Hillary II. Vala, Jamie III. Title.

QH75.S47 2004 333.95 C2004-903813-3

Captus Press Inc.
Units 14 & 15
1600 Steeles Avenue West
Concord, Ontario L4K 4M2
Telephone: (416) 736–5537
Fax: (416) 736–5793
Email: info@captus.com
Internet: www.captus.com

Canada We acknowledge the financial support of the Government of Canada through the Book Publishing Industry Development Program (BPIDP) for our publishing activities.

0 9 8 7 6 5 4 3 2 1
Printed in Canada

Contents

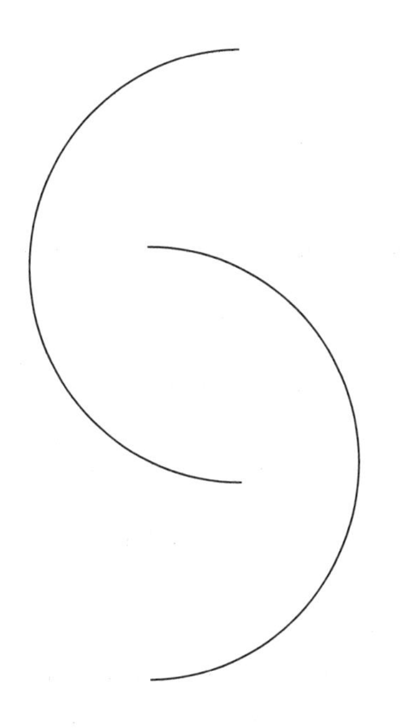

Acknowledgments

Dozens of people have contributed to the accumulated knowledge and perspectives presented here. They have worked for various projects of the Douglas College Institute of Urban Ecology, later part of the Douglas College Centre for Environmental Studies and Urban Ecology. In particular we would like to acknowledge: Mona Keffer, Sheryl Webster, Marty Sulek, Naomi Tabata, Amanda Carr, Maya Levy, Melinda Yong, Nicole Lee, Rob and Renata Hewlett, Kelly Fujibayashi, Tracy Hetherington, Leanne Paris, Katherine Mikes and Jennifer Eakins. This book is a testimony to their creativity and passion for urban biodiversity.

Sponsors for projects that produced information that was subsequently used in the book included: BC Hydro, BC Gas (later Terasen Gas), the Real Estate Foundation of BC, the BC Ministry of Environment, Lands and Parks (later the Ministry of Air, Water and Land Protection), Environment Canada (Environmental Partners Fund, ecoACTION), Canada Trust Friends of the Environment Foundation (later TD Friends of the Environment Foundation), the Habitat Conservation Trust Fund, the Vancouver Foundation, Human Resources Canada (Job Creation Partnershp Program), Canadian Council for Human Resources in the Environment Industry, City of Surrey, City of Coquitlam, VanCity, University of Victoria Environmental Studies Program and the EJLB Foundation. Their ongoing support has enabled us to conduct years of research, restoration and stewardship that have given us the experience to explore biodiversity and nature's services in an urban context.

Nicole Lee contributed to the section on indicator species and related topics. The discussion on cemeteries, golf courses and airports was based on work by Rosalie Aguilar. Line graphs are by Gavin Schaefer unless otherwise noted. Travis Deeter prepared the list of organizations in Appendix A. Anny Schaefer provided creative and editorial comments. Naomi Tabata checked an earlier version for typographical errors.

Photo credits: The following people provided photographs with permission for this book: Nicole Lee - mycorrhizal innoculum, people constructing a wattle, and child planting a seedling; Anny Schaefer — Seattle rooftop gardens, ravine cleanup, people hugging giant tree; Mark Fallat — spider; Hillary Rudd — water quality testing; Douglas College Institute of Urban Ecology (photographers unknown) — Green Links group photograph and woman passing nail to child building bird box. All other photographs in the book are by Valentin Schaefer.

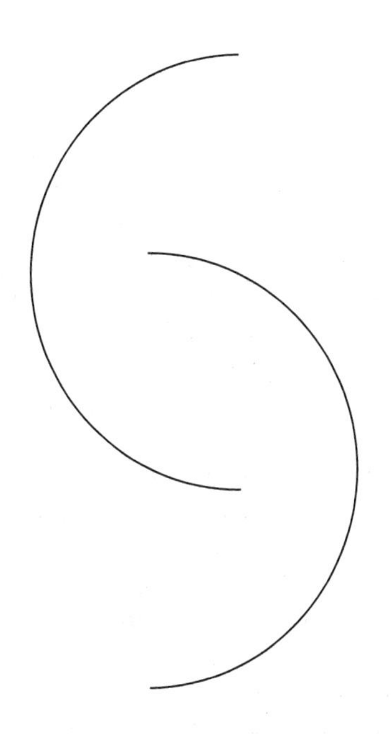

Preface

Human impacts on the planet are considerable. The global population, now in excess of 6 billion, will continue to grow over the next few decades and is expected to eventually stabilize at around 10 billion. Compounding the environmental impacts of our sheer numbers are the lifestyle choices of many developed countries which, if applied throughout the world, would already require three planet Earths to satisfy. We are currently drawing 40% of the planet's gross terrestrial productivity and have directly left our mark on 70–80% of the planet's surface. Indirectly, we affect the entire planet.

Although we are destroying habitat and biodiversity, we ultimately depend on this habitat and biodiversity for survival. This has been highlighted by a collection of papers in 1997's *Nature's Services: Societal Dependence on Natural Ecosystems*, edited by Gretchen Daily. She presents an overview of how Mother Nature looks after us; providing us with food, water, and shelter and cleaning up our messes.

More than 50% of the global population and 90% of North Americans live in cities. Although we might dismiss biodiversity in urban settings, it turns out that it is very important. Cities are often located in biologically important sites such as estuaries and floodplains. They are strategically located on migratory routes and on major breeding grounds for wildlife. We have ignored cities as a biological resource, and there is a great deal that can be done in private and public landscapes for biodiversity.

Urban Biodiversity: Exploring Natural Habitat and its Value in Cities looks at the impacts of urbanization on biodiversity and the value of biodiversity in providing services at both a global and an urban level. We highlight biodiversity in the city with an overview of different habitats and suggest some simple ways individuals can increase biodiversity.

Although humans evolved in natural settings, we have become an urban species. We often get caught up in the material world, yet in our more spiritual moments we find ourselves caring about nature. We acknowledge its intrinsic value, not just its instrumental worth to our comforts and the economy. At such times we have an awakening, a realization that nature is with us in the city. We think of how we can care for nature, not just how it can serve us. We are aware of urban biodiversity and that working to enhance wildlife habitat is part of having a healthy mind, a healthy body, a healthy city and a healthy planet.

About this Book

This book is intended for both a popular audience and as a university textbook. Part 1 begins with a simple overview of basic ecology to define some terms and present some concepts. This is followed with a discussion of biodiversity in the urban context, describing how the urban environment presents a different set of conditions for natural ecosystems, what characterizes urban biodiversity and how the contributions of biodiversity to maintaining the quality of life in a city can be measured.

Part 2 is an overview of the different habitats found in the city. It provides a perspective on the range of habitat types and their differences as natural environments.

Part 3 explores how urban biodiversity is conserved and enhanced. It contains information on individual action, creating backyard and balcony habitats, and actions that are taken by various levels of government and international agreements. It ends with a collection of case studies of actual projects to improve urban biodiversity.

The book is well referenced and contains an extensive bibliography. The appendix lists a number of organizations involved with Urban Ecology that can be useful contacts in pursuing this area of activity yourself. A glossary at the back of the book offers definitions for some of the terminology.

Part 1

Nature in the City: Basic Concepts and Impacts

Some Basic Ecology
Urban Landscapes and Biodiversity
Indicators of Urban Biodiversity
Urban Ecosystem Services
Evaluating Nature's Services in the City

1

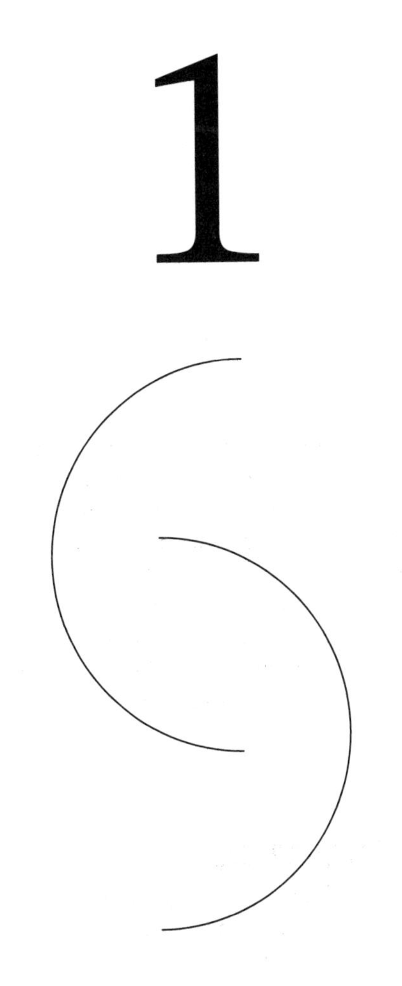

Some Basic Ecology

1

Where do you begin to study nature? It is an immensely broad topic that encompasses many scientific disciplines and frequently leads to questions that remain unanswered. How, then, can you make sense of the complex phenomenon we call nature?

One approach is to look for patterns. This is ecology, the study of the living (biotic) and nonliving (abiotic) environment and the interactions between them. To do this requires some basic vocabulary to refer to nonliving things, to species and to the interactions themselves. This can become technical but everything relates to energy flow and nutrient cycles.

Two Basic Themes

The two basic themes in ecology are energy flow and nutrient cycles. It is for the purpose of obtaining energy and nutrients that organisms interact with each other and with the nonliving world.

Energy Flow

All organisms require energy in order to live. How they obtain this energy, how it is used, and how it is passed on to other organisms is described by food chains and food webs. The ultimate source of energy is sunlight. The energy passes through a series of plants and animals, to be eventually lost as heat to the atmosphere and the universe. There are two types of food chains, grazing food chains and detritus food chains.

- Grazing food chains begin with green plants and go on to herbivores, which feed on these plants, and then the carnivores, which feed on the herbivores and other carnivores.
- Detritus food chains begin with a base of dead organic material. Detritus is dead organic material and bacteria that is eaten by micro-organisms that in

Figure 1.1 Grazing Food Chain

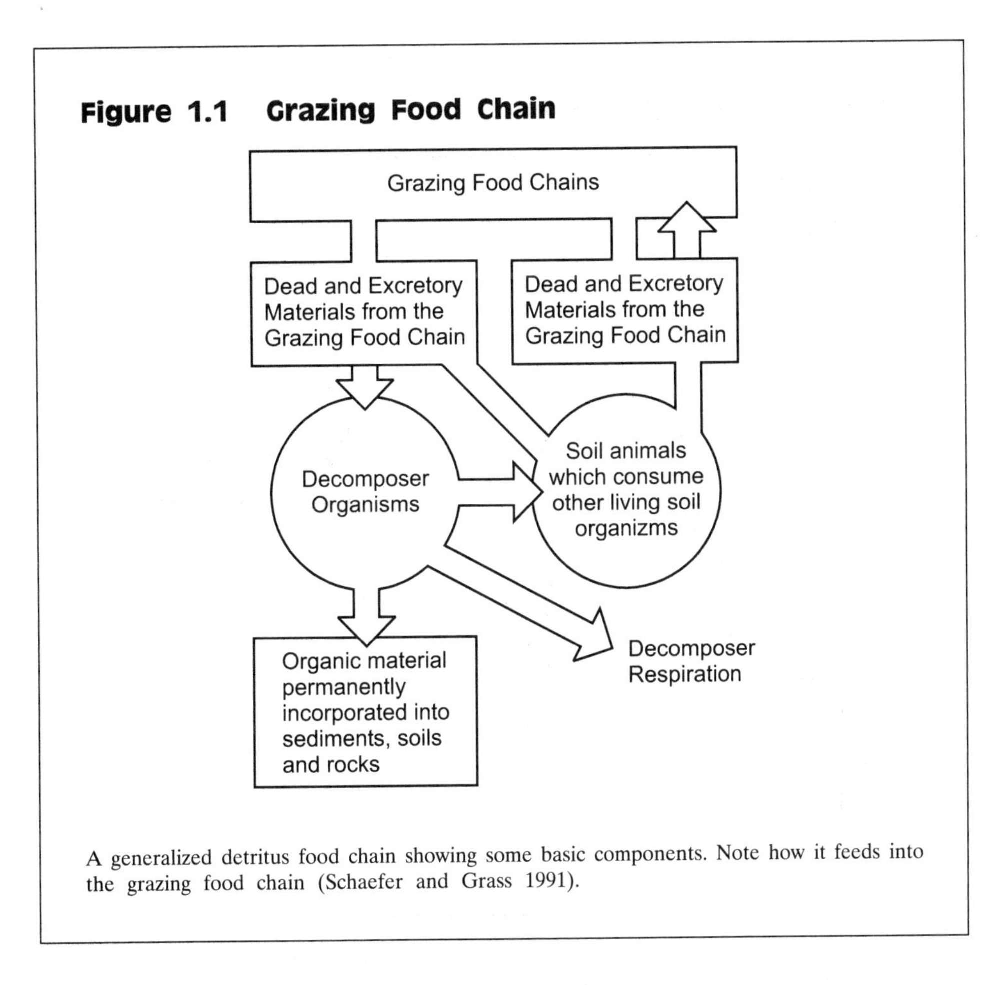

A generalized detritus food chain showing some basic components. Note how it feeds into the grazing food chain (Schaefer and Grass 1991).

turn are fed upon by detritivores. Detritus food chains feed into grazing food chains through producers (Figure 1.1).

Usually many food chains are interconnected. These interconnected food chains are collectively called a food web.

Groups of organisms in a food web that obtain their energy by the same number of steps (levels) on a food chain are said to occupy the same trophic level (the same feeding level). Green plants occupy the first trophic level — the producers (Figure 1.2). The producer is autotrophic (self-feeding) because it can sustain itself with simple nutrients from soil or water and with an external energy source (usually sunlight). The second trophic level consists of the primary consumers that feed upon plants. A consumer is heterotrophic (literally, other feeding as opposed to self-feeding) and obtains both its nutrients and energy from other organisms. The third trophic level is the secondary consumers that are primary carnivores that eat herbivores. The secondary carnivores are the tertiary consumer level, and so on.

Usually there are only five or six trophic levels since only about 10% of the energy at one level is passed on to the next. The other 90% is "lost" in keeping the organism that has it alive. Other energy is unavailable to the next

Figure 1.2 Energy Pyramid

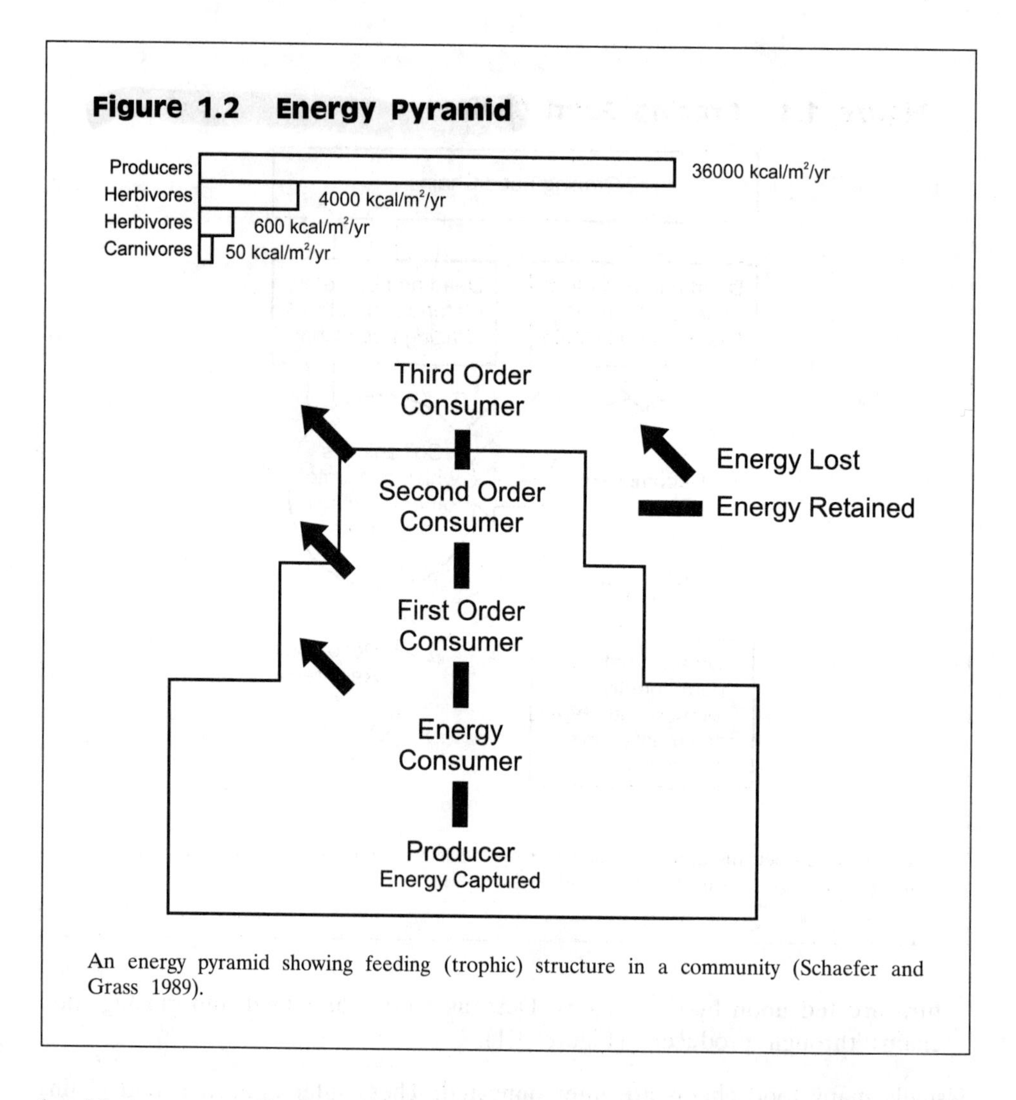

An energy pyramid showing feeding (trophic) structure in a community (Schaefer and Grass 1989).

level because not all of the food can be found and eaten by consumers. It does not take many links in a food chain before it runs out of energy.

Productivity

The total amount of energy originally fixed by plants from sunlight is called *gross primary productivity.* The plants use up some of this energy themselves. What is available to pass on to the next trophic level is called *net primary productivity.*

Nutrient Cycles

The second central theme in ecology is that of nutrient cycles. The basic materials (nutrients) required by organisms to live are obtained from other organisms or the nonliving environment. Unlike energy that flows through a food chain and is then lost, nutrients are constantly recycled. Nutrient cycles include a circuit through plants, animals, decay and soil, or some other physical state, like the Earth or atmosphere.

All organisms require nutrients in order to build the various molecules necessary for life. The major molecules of life, the organic molecules, are carbohydrates, lipids, proteins and nucleic acids. The most common building blocks of these molecules are the elements of carbon, oxygen, nitrogen and phosphorus, although there are many others. These elements are the "nutrients", and they flow through the living and nonliving world. The living components include bacteria, protozoans, fungi, plants and animals. The nonliving components include water, soil, rock and air. Nutrient cycles are sometimes called biogeochemical cycles to reflect the fact that the living and nonliving components are involved, and that the nutrients pass through a number of chemical states. The nitrogen cycle is shown in Figure 1.3.

Figure 1.3 Nitrogen Cycle

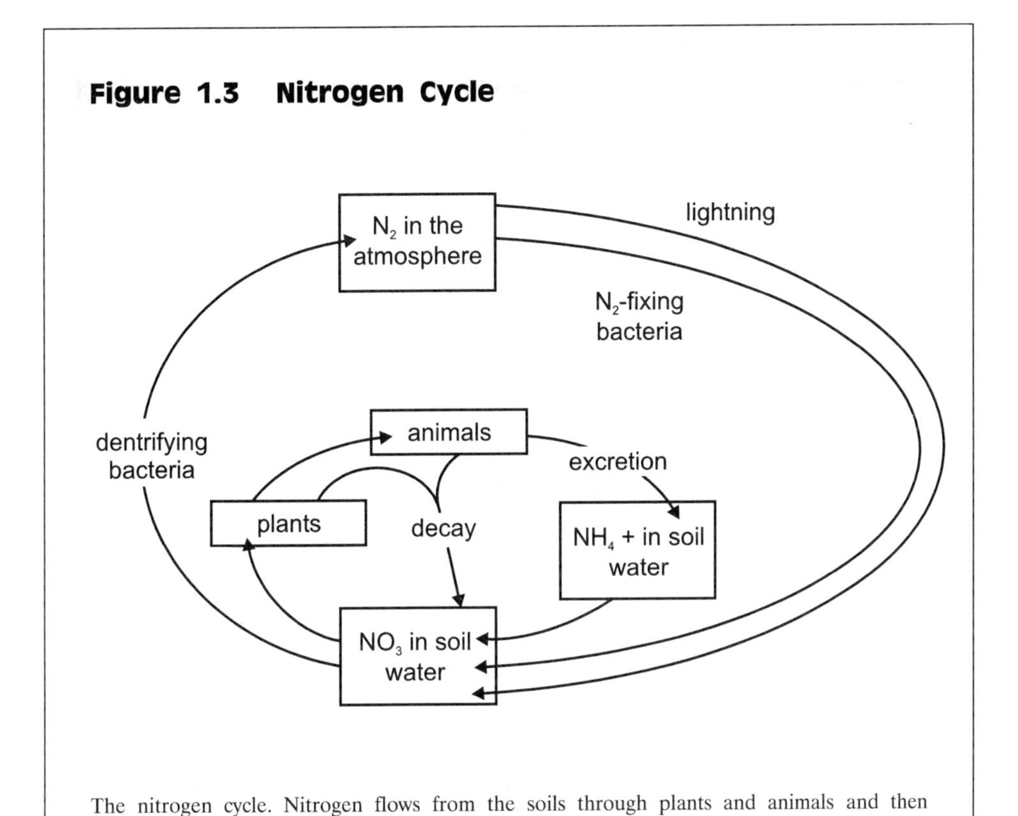

The nitrogen cycle. Nitrogen flows from the soils through plants and animals and then back to the soil again by excretion or decay. Nitrogen also moves into the soil through the action of nitrogen fixing bacteria or lightning, and can return back to the atmosphere by denitrifying bacteria (Schaefer and Grass 1989).

Populations, Communities and Ecosystems

The *ecosystem* is a major functional unit in ecology and literally means an ecological system. It is defined as a system containing living and nonliving components that interact with each other according to a particular set of relationships. In these interactions what is being exchanged, in one form or another, is nutrients or energy.

An ecosystem in turn, is composed of a number of *communities*. A community is simply defined as all organisms (including all species) found in a given area and is not necessarily a distinct functional unit as would be an ecosystem.

Organisms of the same species found in a given area form a *population*.

Ideally, the ecosystem would be self-contained. The only input would be sunlight energy and the only output would be heat, with all life and living processes happening in between the light and heat stages. In reality, there is only one perfect ecosystem and that is the biosphere itself (the biosphere is that part of the planet Earth where life exists, a thin layer extending from the

Figure 1.4 Biomes and Climate

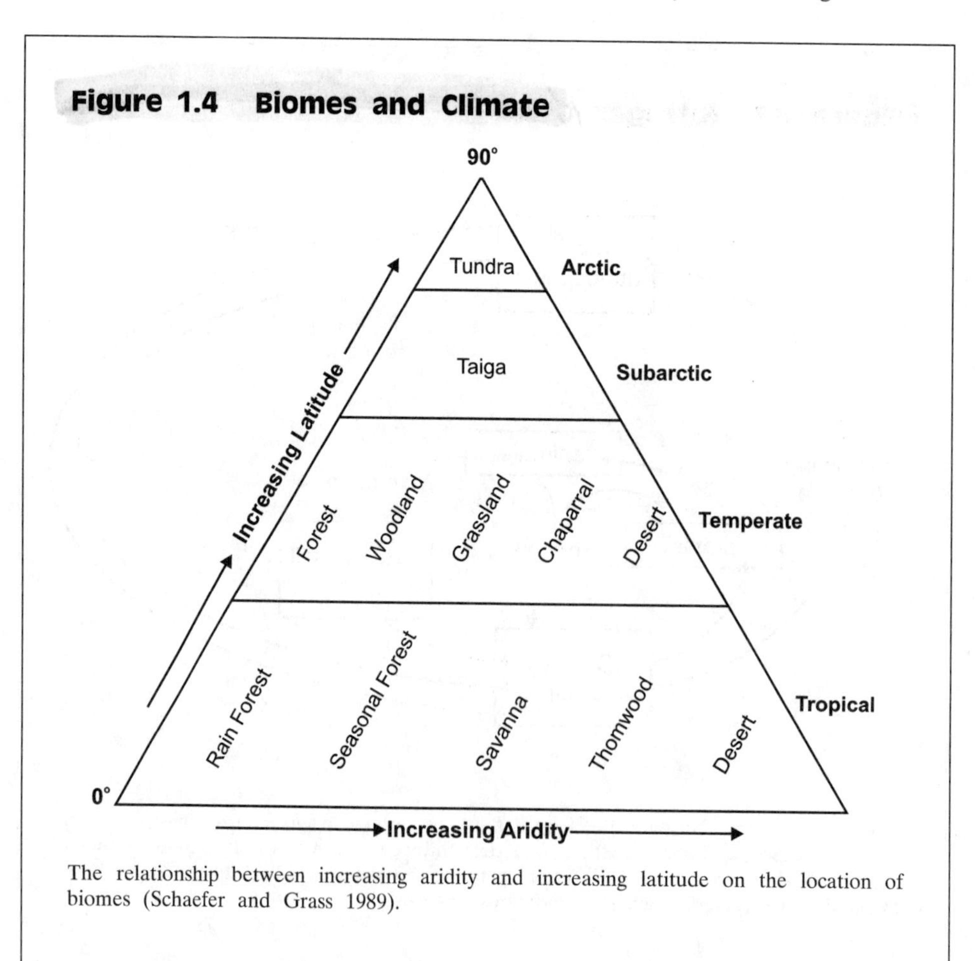

The relationship between increasing aridity and increasing latitude on the location of biomes (Schaefer and Grass 1989).

earth's surface to a few kilometres into the atmosphere). All other ecosystems are collections of communities, including all of the living organisms together with the nonliving environment, forming a relatively self-contained interacting system. It is an arbitrary unit that can be used to describe the planet as a whole, or a system as small as a tidepool (Schaefer and Grass 1989).

Ecosystems can occur on land (terrestrial), or in water (aquatic — both freshwater and marine). The major terrestrial ecosystems are the biomes. The vegetation on our planet forms very large, recognizable associations of plants called biomes that are determined by the climate and soil of an area. Patterns in animal life follow those of the plants. There are 12 major biomes in the world, examples being the tundra, coniferous forest, deciduous forest, grassland and desert.

Ecosystems in freshwater can be standing water types such as ponds and lakes, or flowing water, such as streams, rivers and estuaries.

Examples of marine ecosystems include: intertidal, subtidal, open water or pelagic and deep sea thermal vents.

Succession

Succession refers to the replacement of one community by another in an area, ending with a stable climax community that can maintain itself indefinitely if left undisturbed. Communities are constantly being disturbed. For example, a forest can be destroyed by fire or a hurricane. A community that has established itself over time is the one best suited for the climate and soil conditions of a particular area and is called the climax community.

If the climax community is destroyed it cannot usually replace itself immediately. A number of other communities must first occupy the area, changing the conditions of the site to those more favourable for the climax community. For example, coniferous trees, such as hemlock and cedar, form the climax community in the coniferous forest biome but they cannot establish a community in open sunlight. Other plants must first occupy the site and create the shade that conifers need to grow. In the case of the major terrestrial communities, the climax community is the biome.

Habitats and Niches

A *habitat* is the place in which a plant or animal lives. The term usually refers to the land or water the organism occupies and which also determines its survival. Although often seen as the organism's "address", it may not be that simple. Some habitats are defended; for example, many songbirds do so during breeding season. Some habitats are used infrequently, such as the home range (the area in which an animal moves) of a large mammal, like a grizzly bear, for example, that may need several hundred kilometres of habitat, but can only be in one small part of it at a time. Some habitats are separated in space (migrating animals need habitat in two, often very distant locations) or time (for example, a migrating breeding bird may require berries that are only available in the summer). Some habitats are very defined and discrete, such as a patch of forest, whereas other habitats, such as backyard habitats, are diffuse. The biodiversity benefits of discrete habitats are usually more readily evaluated than those of diffuse habitats.

Habitats can be as small as a bird nest. Some insects that are commensals (use the nest as a home, but do not harm the bird) in the nest with a bird include carpet beetles (*Attagenus pellio* and *Anthrenus scrophulariae*), spider beetles (*Ptinus tectus* and *Niptus holoeucus*) and hide beetles (*Dermestes lardarius* and *D. maculates*). Also found in the nest would be bird parasites, such as the bird flea (*Ceratophyllus gallinae*) and acarid mite (*Glyciphagus domesticus*) (Gilbert 1991).

The *niche* of a species is broader than the habitat in that it encompasses the role the species plays in an ecosystem. The concept of a "niche" has changed over the years. It used to simply describe the habitat of the species (spatial niche), then included its interactions with other species (functional niche), and now describes any conceivable connection between the species and the living and nonliving world (multidimensional or hypervolume niche).

Some fungi are detritivores that break down organic matter: some just cellulose, others lignin as well. Mycorrhizae are fungi that live symbiotically in association with plant roots and help plants to absorb water and nutrients. Pictured here is an innoculum of mycorrhiza being placed in the ground for a planting of tall Oregon grape to try to help it survive in a gravel pit.

Interactions between Species or Organisms

The interaction between species is called symbiosis (living together). In feeding relationships this takes the form of predation, where the predator kills its prey, or parasitism, where the parasite feeds off its host but doesn't usually kill it.

In relationships where species are seeking nutrients and energy from the nonliving world, the species can either compete or cooperate. In competition, each species does less well in the presence of the other and would be better off if the other one wasn't there. Both species are harmed in some way because they lose resources and energy to their competitors.

The cooperating relationships include mutualism, where both participants benefit, and commensalism, where one benefits and the other is unaffected. A good example of mutualism is the lichen, an organism composed of an alga and a fungus. The alga undergoes photosynthesis and obtains energy from the sun, the fungus provides nutrients from the rock or dead material and some moisture. An example of commensalism is a licorice fern growing on a tree trunk, gaining access to more sunlight and not hurting the tree.

Biodiversity

Biodiversity is defined as biological diversity, or the variety of life in an area. Taxonomists (scientists who classify organisms, placing them into groups and categories on the basis of similarities and differences) recognize five kingdoms of life with the following numbers of species (Wilson 1992):

- **Kingdom Monera** — Bacteria (5,000 species)
- **Kingdom Protista** — Single-celled with a nucleus (31,000 species)
- **Kingdom Fungi** — Mushrooms, mildews (69,000 species)
- **Kingdom Plantae** — Plants (250,000 species)
- **Kingdom Animalia** — Animals (1,032,000 species)

The actual number of species on Earth is not known. About 1.5 million have been classified so far, half of them insects (Wilson 1992). Estimates of the total number of species vary from 10–100 million, with 30 million being a good guess at the moment.

However, there is more to biodiversity than just the number of species. There are four levels of biodiversity:

- **Ecosystem** — assemblages of communities of different species form complex interrelationships
- **Species** — each individual species is unique
- **Population** — groups of individuals within species may have distinctive features that form the basis of short-term microevolution, adapting the species to change
- **Individual** — genetic differences between individuals contribute to the makeup of populations

Within these levels are also structural and functional attributes (Noss 1990) that add to its complexity. The genetic composition of certain populations that occur, say, at the edge of a species range may be more important to the long-term survival of a species because of the intense selection pressures in populations found in areas of extreme environmental conditions for the species. Most large cities in Canada are along the southern border of the country, an area corresponding to the northernmost limits of the ranges of many North American species of plants and animals. The genetic makeup of urban populations for these species may be of special importance (Mosquin 2000).

When discussing "biodiversity" we are actually referring to all levels: ecosystems, species, populations, individuals, parts of individuals. Each level is determined by heredity — the genetic makeup of the cells. What they all have in common is that they are referring to "differences". It is these differences that provide the raw material for survival, the ability to adapt to change in the short term, and to evolve in the long term.

It is a general rule with populations, communities and ecosystems that the greater the biodiversity, the greater the ability to adapt to change. Reduced biodiversity leaves an ecosystem unhealthy and vulnerable. Species become extinct and productivity is reduced. Urban biodiversity in particular is faced with serious challenges of habitat loss and disturbances to air, water and soil. Nevertheless, urban biodiversity is of global importance to the biosphere, and the need for biodiversity couldn't be greater.

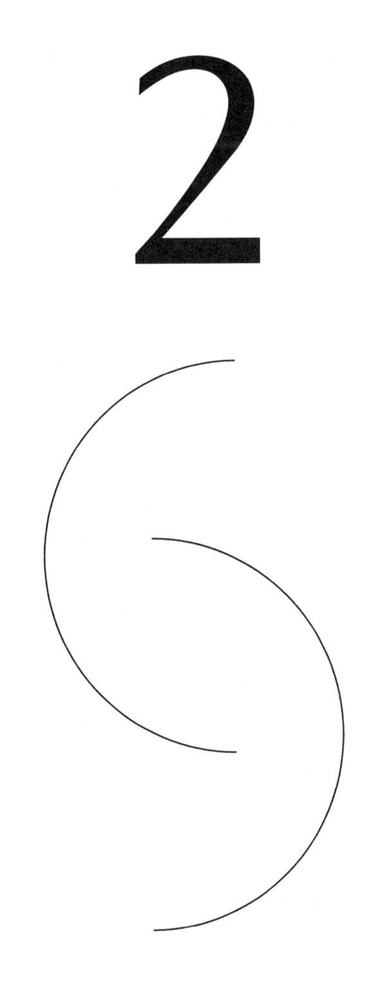

2

Urban Landscape and Urbanbiodiversity

The year 2000 was, for many, a time to reflect on our history and think of our future. It was a time to celebrate the information age, with the cellphone potentially giving everyone access to each other, anywhere in the world, without the need for a wire connection. It was also a time to re-evaluate the loss of spirituality (life and energy within the mind and mysterious in some way), that high technology seems to have precipitated. Many people now seek to regain their spiritual sides by turning to familiar Western religions, to the mystical appeal of ancient traditions still honoured in Eastern religions and First Nations cultures, and to nature. Some ecologists try to apply a more spiritual approach to their analysis of natural systems and human impacts (see Devall and Sessions 1985).

The move to embrace nature today is not a rejection of capitalism, consumerism and the city, as was perhaps the case in the 1960s and 1970s. It does not promote finding freedom on 50 acres (ac) in the wilderness or country. Instead, it is a movement to embrace nature in our lives in the city. It is a movement to discover nature in the city, to celebrate the birds, butterflies, bats, and bees with whom we share our living space. It is also a movement of environmental stewardship that not only supports the conservation of large natural areas and a global awareness, but also encourages environmental stewardship of our own backyards, creating wildlife habitat in small places and enhancing the streams, fields, and forests that remain in the city.

Cities are a relatively new development on the planet Earth. They first appeared along the Nile, Tigris, and Euphrates Rivers of Egypt and the Middle East about 5,000 years ago. They evolved from benefits of agricultural surpluses, expanding populations, and a sense of common interests. They grow because of a migration of people into the city from rural areas, because of the natural increase of people already living in cities and because they sprawl, leading to a reclassification of rural areas as cities.

Cities with 1 million or more people did not appear until Peking, China (now Beijing, China) reached that milestone in about the year 1800. There were 16 cities in the "million club" by 1900, and in 1980 there were 235. Predictions are that by the year 2025 there will be 135 cities with 4 million or more people, which at that time will also be about 25% of the world's population (Good 1990).

By 1900, some 14% of the world's population, or about 200 million people, were living in cities. This rose to 600 million in 1950, 2 billion in 1986, and 3 billion in 2000. The current growth rate of cities is 2.5% per year, leading to a doubling of the urban population in 28 years (Brown and Jacobson 1986).

Urbanization has been more rapid in highly developed countries that were the first to industrialize. These include the United States, Canada, Japan, Australia, and the countries of western Europe and Scandinavia. In these places about 80–90% of the populations live in cities. The percentage of people living in cities has been about 65% in moderately developed countries, such as those of Latin America (Mexico, Central and South America), north and west Africa and east Asia. The same is true for China, which is somewhat less developed, and the lesser developed countries of India and east and central Africa.

The modern urban landscape, with its extensive blacktop and lawns only became common after World War II due to the prevalence of the automobile and rapid population growth. The mass production of automobiles enabled people to live further from work but brought with it a dependence upon the vehicles. In the United States about 90% of people drive to work compared to only 40% in Europe (Brown and Jacobson 1986). Urban impacts on the natural landscape are great, extending beyond the automobile to the suburban culture of lawns and barbecues. For example, about 2% of the entire land base of the United States is composed of lawns (Garber 1987); lawns have reduced biodiversity, and require watering and cutting.

Our population growth rate has begun to decline during the past two decades. Still, the current global age structure and population growth rate indicate that our population should stabilize at about 10 billion people around the year 2050. What is perhaps more important at the moment than population size, especially in industrialized developed countries, in terms of our impact on the environment, is our lifestyles. We consume far more resources than the environment can provide locally. We exceed our local carrying capacities and the shortfall is "appropriated" from other areas of the planet.

One concept used to estimate the amount of land it takes to support our needs and lifestyles is the ecological footprint (Wackernagel and Rees 1996). The estimate includes the land needed to grow our food, materials to make our clothes and grains to produce the fuels to drive our cars. Wackernagel and Rees calculated that since 1900, the per capita land needed to support populations in industrialized nations has increased from 2.5 acres, about 1 hectare, to 7 acres, about 2.7 hectares. At the same time, the per capita productive land area has decreased from 14 acres (5.4 ha) to 3.7 acres (1.4 ha).

Wackernagel and Rees calculated the amount of land needed to support the current lifestyles of the 2.1 million people of Greater Vancouver. The actual land area for Greater Vancouver is 400,000 hectares, and the population needs 8.1 million hectares. The 7.7 million hectares of extra productivity is appropriated from other areas of the planet, or from historical reserves of fossil fuels.

A similar concept promoted in Europe is the "ecological rucksack." It is intended to bring to light the hidden inventory of materials, resources and impacts involved in producing a product. For example, to produce a quart of Florida orange juice requires two quarts of gasoline and a thousand quarts of water. Another similar analysis is that of the "food mile." It is not uncommon for the food on our plates to have come from places 1,000 km away. The food mile is one way to register these distances and compare the environmental impacts of imported foods with those grown more locally.

What the ecological footprint has demonstrated is that even today, for everyone to enjoy the North American lifestyle, we would need three planet Earths. With 10 billion people, five planets would be needed. There will have to be some lifestyle changes to accommodate this growth. Hopefully they will lead to a better quality of life, even though our consumption of material goods will decrease.

Our cities will continue to grow in size and numbers. Like it or not, our experience with nature will more and more be an urban experience. Urban environments need to be included in a comprehensive global biodiversity conservation strategy.

Population Growth and the City

Before the agricultural revolution about 10,000 years ago, the world's population was only a few million. With the advent of agriculture this number rose to a few hundred million people, where it remained until the beginning of the nineteenth century. The Industrial Revolution brought about another population explosion. By 1830 the population increased to 1 billion. By 1930, only 100 years later, the population had doubled to 2 billion. With new medical advances it only took 30 years to reach 3 billion and, by 1975, it was already 4 billion. In 1987 — only 12 years later — the world's population reached 5 billion. In 2000 we crossed a threshold where over 50% of the world's more than 6 billion people now live in cities.

In the natural world, the city would be similar to other animal colonies and would be no more a problem for the biosphere than a hive of bees or a termite mound. The growth and activities of our colony (groups of people interacting to create an environment of mutual benefit) would be bounded by the limits of nature, as are colonies of bees and termites.

However, cities are not constrained by natural processes, as is the case for other colonial animals. The natural checks and balances are distorted by our use of machines and technologies. Cities require huge injections of energy, food and consumer goods far in excess of what the land base of the city can naturally provide. Cities produce pollution and liquid and solid wastes in quantities far too great for the land base to absorb.

Cities have some basic requirements to provide people with food and shelter. Added to this are nonessential items required to support our standard of living. Cities are a product of satisfying both our needs and wants. Coming into the city are food and water, fuel and energy, processed goods, pulp and timber, and additional building materials.

The natural world has a "circular metabolism", where nutrients are recycled. Plant and animal wastes are produced on a scale that can be handled by the organisms (detritivores, fungi, bacteria) that decompose them into nutrients available to other organisms. Energy is not recycled, but is on a scale matched by that which can be provided for by the ultimate source of energy, the sun, either locally or over small distances.

So far cities have been modeled on a "linear metabolism" (Girardet 1992). Nutrients are not recycled and energy consumption has exceeded the natural supply of the sun by exploiting a historical build up in the form of fossil fuels. The consumption of these fossil fuels has added to the unnatural ecosystem outputs associated with cities such as nitrogen dioxide, sulphur dioxide, ozone, and particulate matter.

The linear metabolism reaches far beyond city limits. The actual area of land a city occupies is nowhere near what is required to satisfy its residents' needs and wants. Food and other materials must be brought to cities over long distances. This creates long supply lines for food and raw materials resulting in more pollution from the machinery and the burning of fossil fuels this entails. The average energy kilowatt-hour moves about 200 miles from production to consumption; the average food molecule moves about 1,000 miles. A city of 1 million imports 2,000 tons of food, 10,000 tons of fuel and 620,000 tons of water daily, resulting in a million tons of garbage and 400,000 tons of human waste annually (Morris 1990). In addition to being transported, food also requires more processing, packaging and refrigeration to survive transport.

We must also look at the land, which is being used to produce the food and goods of a city in more distant regions. Wackernagel and Rees (1996) did an analysis for Vancouver, British Columbia of what they referred to as the appropriated carrying capacity of the city. Greater Vancouver's population of 2 million lives on an area of 400,000 hectares. However, Wackernagel and Rees determined that the land requirement needed to support the needs and wants of the population was in fact 8.1 million hectares. The city requires 7.7 million hectares of land outside the city limits to support its residents.

Instead of being in harmony with the natural world, cities generate an output of sewage (liquid waste), garbage (solid waste), and a whole host of chemicals that pollute air, soil, and water. Cities do not have to be like this. There are visions and plans to convert a city's linear metabolism to a circular one. In this scenario organic waste is composted, other solid wastes are recycled, there is a reduction in fossil fuel use through lifestyle changes and the remaining pollutants are processed naturally through increased vegetation and green spaces.

The way cities are built to accommodate urban nature and green space is more important now than ever before. Some wildlife benefit from urban development, but most are pushed out of their natural range because they are unable to survive the habitat destruction and the artificial-built environment (Garber 1987). From an economic perspective, when we lose green space we risk losing the services it provides, like clean drinking water. We must then replace these services with technology that is often very expensive (like a water purification plant).

Since the 1970s there has been an increasing interest in nature in the city. This is partly because we are aware that green space and wildlife within the city greatly add to the quality of life and livebility of a region. Trees shade buildings on hot days and warm them on cool ones, protect buildings from wind, and clean the air of harmful pollutants. As people become more interested and aware of nature in the city, they also recognize how the natural and artificial components of the urban landscape interact to provide ecosystem services essential to human and wildlife survival. Having a natural presence in the city serves as a constant reminder that we are part of an ecosystem and that we must live our lives with a sensitivity to the limits of its capacity to deal with our waste and to supply us with food and the basis for material wealth.

Naturescaping the City

There has been an evolution in how we landscape residential gardens and commercial properties over the last century. Traditionally there was no attempt to save natural environments or to acknowledge the biological value of urban flora and fauna. Planners and landscape architects historically used a gardenscape approach that involved controlled planting, emphasizing visual aesthetics and design. This typically involved using non-native plants that look nice, but added little to the ecological value of the region. The plantings also required intensive maintenance and watering to remain visually appealing.

The city offers another highly artificial landscape — the containers of concrete, brick, asphalt, glass, and other synthetic materials. The plantings are small and even more uniform than in the gardenscape style and have even less ecological value.

A more recent approach to urban landscaping is the ecological style, sometimes called naturescaping or wildscaping. Perhaps the first example of this approach used on a wide scale in North America was in Texas with the State's Wildscapes Program and reduced cutting of highway roadsides. Features of the natural local environments are used, and the natural elements are allowed to function with no (or very little) maintenance. In addition to retaining natural habitats in development, this style also encourages the use of native vegetation in landscaping, thereby encouraging wildlife in the city. The result is a landscape that adds to the character of a city. Using plants from the local bioregion distinguishes the city from others in the international community.

Characteristics of Natural Environments in Cities

The development of a city and the mobility of its inhabitants have profound implications for the composition of natural habitats. Green spaces have typically been an afterthought and are usually small and isolated from each other. The high degree of disturbance and removal of a significant amount of natural vegetation mean that once flourishing ecosystems no longer function as well as they once did.

Through development many changes occur in cities, including the introduction of other non-native species, increase in temperature (urban heat island), dramatic changes in hydrology, and a reduction in species that are sensitive to development.

Changes in Biodiversity

Although cities have less ecosystem diversity than surrounding natural areas, there are a high number of introduced plants concentrated in cities, making the species diversity artificially high. There are more species of plants within a city than within a comparable area of a natural ecosystem because of the combination of native and introduced species. Suburban areas especially tend to have a biodiversity "peak" in species numbers compared to more natural rural areas or preserves (McKinney 2002). This is explained by the intermediate disturbance hypothesis: initial human disturbance from suburban sprawl is relatively low where we find only a few housing subdivisions in a large matrix of agricultural land or natural areas. The habitats in this heterogeneous mixture still retain much of their biodiversity, resulting in a net gain of species due to urban development.

In fact there are some species that only exist in cultivation and no longer occur in nature. The ginkgo tree is a famous example. It was spared extinction from development because it was cultivated in Chinese and Japanese monasteries. There are also many introduced plant species that become invasive when introduced to a new environment. In the Lower Mainland of British Columbia, Scotch broom (*Cytisus scoparius*), Himalayan blackberry (*Rubus discolor*), and purple loosestrife (*Lythrum salicaria*) are three examples of species that have left behind the checks and balances of their natural habitats and have become invasive in their new habitat. Smith et al. (2000) document the rapid spread of the small Asian pistache tree (*Pistacia chinensis*) in Armidale, New South Wales, Australia, since the 1940s through a combination of extensive deliberate plantings to beautify neighbourhoods with its autumn colours, and efficient avian seed dispersers.

The diversity of native species of birds and mammals decreases in cities since many essential components of their habitats are lost through urban development. For example, although there are over 350 bird species found in the Greater Vancouver region of British Columbia, less than 50 species are commonly found in the city. The most abundant of these 50 species are those that have been introduced, like the European Starling (*Sturnus vulgaris*), which occurs in flocks of thousands of birds.

Not only is bird life in the city reduced by human disturbance; sometimes other bird species that have been introduced and do well in urban environments displace native species. For example, bluebirds *(Sialia sp.)* are thought to have been pushed out by European Starlings; both birds are hole-nesters (nest in cavities), and the larger, more aggressive starlings displaced bluebirds, even though they would otherwise be expected to do well in urban areas.

Microhabitats

In a natural ecosystem vegetation creates certain types of microhabitats, such as a rotting log, a sword fern herbaceous layer, leaf mold, a vine maple and elderberry understorey. In urban environments the number of microhabitats may actually be greater than in natural ecosystems in a comparable area, although each one may not be as abundant. With the presence of buildings, natural habitats of native vegetation, and a large number of introduced plant species, there can be a broad spectrum of microhabitats in a small area. There are opportunities for species in urban environments that may not exist in more natural settings. Buildings, ponds, street trees, open spaces, and pavement are just some of the microhabitats that exist in urban areas. Examples of birds using alternative habitats are the Chimney Swift (*Chaetura pelagica*) and Barn Swallow (*Hirundo rustica*). Both are cavity nesters and will nest in buildings instead of trees since there are more buildings than tree cavities in cities (Garber 1987).

In this broad assemblage of microhabitats, altered interactions between species and a high influx of introduced species, it can be difficult to focus on what may exist as a reasonable unit defining "habitat". As mentioned earlier, it may be sufficient to merely recognize the biome in which the city occurs, and the larger natural ecosystems around the city.

However, such a broad consideration of natural habitats fails to capture the value of individual representatives of species, or small communities, that perhaps more often are all that represents nature in large parts of a city. Such areas are especially valuable to people's lifestyles because they exist in the built community, and can be enjoyed on a daily basis.

Disturbances to Urban Landscapes

Disturbances in the urban landscape range from natural to human induced. Natural disturbances include floods, landslides, and windstorms. Natural disturbance regimes can cause a significant amount of damage to the urban landscape, but human-caused or unnatural disturbances are much more common and can often be more detrimental to ecosystems found in urban areas.

Urban landscapes are often characterized by development: land being cleared to make way for new buildings. Developed lands and green space are further disturbed through weeding, mowing, and pesticides. When land is cleared for development it no longer provides the components of habitat required for survival. Disturbances in urban landscapes also favour introduced species. Native species are prevented from re-establishing themselves and, in the absence of competition from them, the alien plants thrive. Native vegetation that does become established usually consists primarily of pioneer species. A natural habitat in a city would not be expected to remain undisturbed long enough to form a mature climax community.

Disturbances also have an effect on the interactions between native plant and animal species. In natural ecosystems native plants are often restricted to their optimal habitats by competition from other native species. In disturbed environments some species may find that their competitors are no longer present, enabling them to occupy more marginal environments. An example here is the horsetail that does best in moist habitats. However, in disturbed sites in the absence of its competitors it may be found in drier areas; horsetails are a common sight in vacant lots.

Some Urban Impacts on Biodiversity

Murphy (1988) points out that urban areas are synonymous with a loss of biodiversity. A number of forces are at work besides the direct removal of natural habitats. Reductions in biodiversity can also occur from airborne and water-borne pollutants, over-drafting of local aquifers, and changes in groundwater flows caused by storm drains. Pets and livestock may prey on native animals and spread diseases and parasites among more vulnerable native species. Introduced animals such as the European Starling (*Sturnus vulgaris*), House Sparrow, Norway rat (*Rattus norvegicus*), and house mouse (*Mus musculus*) may become more abundant than native species. Introduced plant species such as the Scotch broom and purple loosestrife (*Lythrum salicaria*) can out-compete and displace the local flora. All these factors can disrupt natural ecosystem functions or cause them to collapse.

The natural world is often not considered by engineers and planners when designing a city or in the downtown core of cities otherwise proud of their natural environment.

Habitat Destruction

Urban environments suffer tremendous losses in biodiversity directly through a loss of natural habitat. Twenty to 30% of the land surface is commonly paved, and buildings cover much of the remainder. Also, native plant species are removed and replaced with exotics. These two major factors cause a loss of ecosystem diversity, and result in extensive local extinctions of native species (see Thomas 1984; Heath 1981).

Liquid Waste

Other major environmental impacts of cities come from the liquid and solid wastes they generate. The amount of waste that each person generates (Table 2.1) can collectively add up to large ecosystem impacts.

Table 2.1 Daily Per Capita Inputs and Outputs

Average Daily Impact of a United States Resident	
INPUT:	568 L water
	1.5 kg food
	7.1 kg fossil fuel
OUTPUT:	454 L sewage
	1.5 kg refuse
	0.6 kg air pollutants

Most of the volume of sewage comes from showers and toilet flushes. Liquid waste problems come from the large volume of organic material and possible toxins from toilets and kitchen sinks. The organics in human waste can ordinarily contribute to the recycling of materials, returning nutrients to the biosphere. However, they are produced in such large concentrations in cities that the oxygen consumed by bacteria to decompose the wastes in the receiving waters is an issue. The wastes also act as fertilizer for algae and higher plants, creating even more organic matter and leading to eutrophication of shallow water bodies. This biological oxygen demand can seriously deplete the amount of oxygen available to fish and kill them.

Clearing land for development is the first single largest contributor to a loss of biodiversity.

Sewage treatment generally involves three levels:

- Primary Treatment: This involves allowing sewage to stay in settling ponds for a period of time that enables about 50% of the organic material to settle out as a sludge. This sludge is collected and composted or disposed. The sewage remaining after primary treatment is called effluent, and is often then chlorinated to kill pathogens and then discharged into a receiving water (usually a river or the ocean). Most cities now have primary sewage treatment.
- Secondary Treatment: The effluent from primary treatment can go to secondary treatment. In this case the sewage is placed in contact with bacteria that can break down most of the remaining organic material. Primary and secondary treatment combined remove 95–99% of the oxygen demand. Alternative

treatments include the use of wetlands to decompose sewage, but they are still only practical for smaller urban areas.

- Tertiary Treatment: This involves removal or neutralization of one toxin (such as a heavy metal) or nutrient (such as phosphates) at a time; different treatments are required to target different specific toxins or problem chemicals.
- Source Control: Liquid waste can also contain toxic chemicals from household sinks or from industry that can be reduced by source control. Source control involves eliminating the need for the toxic chemical in the first place through alternate processes and technologies, or collecting the waste at the location where it is generated so it can be dealt with in some other way (pretreatment) before disposing of it into the sanitary storm drain system.

An additional problem with liquid waste is that many cities historically had at least some of the sanitary sewer drains interconnected with the storm drains that collect rainwater from city streets. Storm drains typically empty into nearby creeks and rivers. Times of heavy rainfall exceed the capacity of the interconnected pipes to handle the flows, in which case sewage might back up into the creeks and rivers. In these situations the pathogens, primarily *Eschericheri coli* bacteria in fecal coliforms, can pose serious health risks to people and wildlife.

In Canada, 78% of the population was served by secondary or tertiary sewage treatment in 1999, compared to only 56% in 1983 (BC Ministry of Water, Land and Air Protection 2002).

Solid Waste

There are several ways of disposing of solid wastes (garbage). These are:

1. Landfills: garbage is covered
2. Dumps: garbage is not covered
3. Baling: garbage is compressed into bales before landfilling
4. Incineration: garbage is burned at high heat
5. Recycling: garbage is collected and remanufactured into other products

Solid waste creates several environmental problems. If garbage is landfilled (whereby garbage is buried), it takes up valuable land close to cities. The largest landfill in the world is on Staten Island in New York; called Fresh Kills, the landfill was already 3,000 acres (1,200 ha) in 1989 (Goldstein and Izeman 1990). A landfill can generate a toxic leachate formed by water percolating through the garbage, picking up toxic chemicals. This leachate eventually reaches the groundwater and surrounding waterways, poisoning aquatic wildlife. This leachate can be collected and disposed of in a safe manner if the landfill has an impervious liner of clay or plastic; however, this is not usually the case. The garbage also produces gas (methane) and odour (hydrogen sulphide). If the garbage is burned it creates problems with air pollution and the remaining ash in the incinerator (about 20% of the original volume of garbage) contains the concentrated toxins. The ash is often considered to be toxic waste.

Solid waste can be largely reduced by composting yard waste, reducing the amount of packaging, recycling paper, plastics and metals, having a deposit sys-

Table 2.2 Types of Solid Waste

Solid Waste Source	Municipal	Commercial
Paper Fibre	39	46
Glass, Metals, Plastics	18	17
Organics	33	25
Miscellaneous	10	12

Source: GVRD, 1991

tem on beverage containers, and encouraging the reuse of materials. Most garbage could be dealt with in this manner (see Table 2.2).

Typical sources of solid waste are residential (33%), commercial (22%), industrial (15%) and construction and miscellaneous (40%) (Environment Canada 1991). Landfilling is still the major method to dispose of wastes, and it is increasingly difficult to find sites. In 2003 the City of Toronto trucked most of its garbage to a landfill site in the United States, while Vancouver trucked about 10% of its garbage to Cache Creek.

There have been great strides made in reducing the amount of solid waste produced, and in recycling the solid waste remaining. The rate of solid waste production was 1.8 kg/person/day in Canada in 1988. The Canadian Council of Ministers of the Environment sought to reduce the amount of garbage sent for disposal per person by 50% (Environment Canada 1991). The amount of material recycled through Ontario's blue box program tripled from 1988–1991.

In the Georgia Basin-Puget Sound Ecosystem of British Columbia and Washington State, the amount of solid waste generated per person dropped from 1,291 kg (2,840 lbs) in 1990 to 1,109 kg (2,449 lbs) in 1999, of which 485 kg (1,107 lbs), or 43% were recycled (Georgia Basin Ecosystem Initiative 2002). The reduced amount of waste and recycling together resulted in 32% less solid waste overall since 1990. In 1999, 8.3 billion kg (18 billion lbs) of waste were generated in 1998 in the region. Of this, 63% was landfilled or burned and 37% was recycled.

Roads and Impervious Surfaces

Automobiles also have an impact on the urban environment by generating air pollution, depositing toxic petroleum chemicals on the ground that eventually run off into streams or rivers, and by the extensive blacktop used to construct roads. The effects on urban streams include an increase in oxygen demand, suspended sediments, hydrocarbons and metals such as lead, zinc, copper, cadmium chromium and nickel from engine blocks and brake linings (Lenat and Crawford 1994).

In addition to an average of 30% of the landscape being lost to roads and parking, blacktop represents a large impervious surface in the watersheds of streams, rivers, and lakes in and around cities. Rainwater can no longer percolate into the ground so not as much water is stored in the ground to recharge streams during the summer months; this kills fish. Similarly, when it does rain, the water over these large impervious areas is all collected into storm drains that connect with others, eventually discharging into surrounding streams. In addition to increased volumes, the water is warmer than it would be naturally because it is heated by the warm black pavement. Impervious surface cover of 10–20% in a watershed doubles the amount of runoff, 35–50% triples the amount of runoff and a cover of 75–100% will result in a runoff five times or greater (Arnold and Gibbons 1996).

The large volumes of water floods creeks, which causes erosion of the banks and siltation, which leads to fish being killed. Cars also leak oil, gas, and radiator fluid, leave worn tire particles containing nickel and cadmium on the road, blow exhaust onto the road leaving toxic residues, all of which contributes to pollution from urban runoff, which is also lethal to fish. Road salt can have an impact on the distribution and abundance of many species of plants and animals (especially amphibians).

Roads and Other Impacts

Roads have additional impacts (after Forman and Alexander 1998), such as:

- Barriers to movement: The width of a road, vehicle use and speed and the spatial patterns of the roads erect barriers to the movement and dispersal of

Impervious surfaces caused by roads have major impacts on surface waters such as streams and ponds. They cause flooding in these waterways during rain events and contribute to droughts in summer because of a lack of groundwater recharge. Pollution from road s degrades water quality.

many animals, or create levels of disturbance that discourage their presence. Roads can subdivide populations affecting their demographics and possibly result in genetic consequences. Carabid beetles and wolf spiders may not cross roads as narrow as 2.5 m wide. The likelihood of small mammals crossing lightly travelled roads 6–15 m wide may be reduced to less than 10% of that of normal movements. A number of measures are available to help wildlife cross roads. Various tunnel designs have been successful in helping wildlife such as frogs, salamanders, turtles and badgers. Along the Trans Canada Highway, larger mammals can pass underneath through arched culverts, boxed culverts and open-span bridges.

- Addition of invasive species: Introductions of exotic woody species along roadsides to reduce erosion, reduce headlight glare or control snow can naturalize and compete with native vegetation.
- Dispersal of invasive plants: Seeds are carried and deposited by cars, or are spread by air turbulence from vehicles. For example, along a highway in New York state the short-distance spread of purple loosestrife occurred along roadside ditches and through culverts.
- Nutrient loading: Nutrients from roadside management and NOx from car exhaust elevated nitrogen levels sufficiently to change vegetation next to a highway for a distance of 100–200m.
- Roadkills: It is estimated that 1 million vertebrates are killed per day in the United States. The highest number of roadkills are probably next to wetlands where one study found more than 625 snakes and 1,700 frogs were killed annually per kilometre. Most roadkill rates do not significantly impact populations, although there are exceptions. For example, the Florida panther (*Felis concolor coryi*) had an annual mortality of 10% of its population from roadkill before 1991. The Western Transportation Institute of Montana State University has designed a sensor to detect larger mammals on a road that would send a signal to a warning sign and flashing lights for motorists that there was wildlife on the road ahead.
- Traffic noise: In one study in the Netherlands, 60% of bird species had reduced numbers in habitat adjacent to roads. The effect-distances (the distance from the road with measurable population reductions) was greatest for grassland species, least for coniferous species and intermediate for birds in deciduous woods. Vegetation in thick strips along with landforms or solid barriers can reduce highway noise by 6–15 decibels (McPherson et al. 2002). More high frequency sounds are absorbed by plants than low frequency; animals tend to be distressed more with high frequency sounds (Miller 1997).
- Traffic density: For the most sensitive bird species and for an average vehicle speed of 120 km/h, the effect-distances through woodlands were 305 m for 10,000 vehicles per day and 810 m for 50,000 vehicles per day. In a grassland environment the effect-distance increased to 365 m for 10,000 vehicles and 930 m for 50,000 vehicles.
- Road density: Populations of large mammals decrease with increasing road density. For example, the maximum road density for a naturally functioning landscape that can support large predators such as cougars and wolves is 0.6 km/km^2 (1.0 mi/mi^2). Measurements of macro-invertebrate density have demonstrated detrimental effects on aquatic ecosystems in areas where roads cov-

ered 5% or more of a watershed in California. In southeastern Ontario the species richness in wetlands correlated negatively with road distances within 1–2 km of the wetland.

On the other hand, roadsides can offer valuable habitat and be reservoirs of biodiversity in situations where all native vegetation was removed for agriculture or urban development.

Fragmentation

Urban biodiversity is not always lost through the total area of land developed. The distribution of the habitat destroyed has an effect on the biodiversity in the habitat remaining. With fragmentation, larger continuous patches of habitat are broken down into scattered, smaller remnants. The biodiversity in the remnants is determined by principles similar to those applying to islands and their biotic diversity (Harris 1984).

Fragmentation effects may ultimately be just as serious at the level of the microhabitat (Wilcox and Murphy 1985). For example, in cities, it is common to remove understorey plants in parks and housing developments. Understorey vegetation contributes greatly to moderating temperature, humidity, light availability, and wind exposure. It is also an important food source and habitat for many bird species (Dean 1976; Laurie 1979).

Hydrology

Changes in drainage patterns in neighbourhoods also affect biodiversity. For example, increased blacktop in a watershed contributes to floods in creeks during periods of rain as less water percolates into the ground and more is collected into the storm drains emptying into creeks. In the summer, when creeks rely on groundwater flows, water levels may drop because the reduced infiltration due to blacktop has lowered groundwater supplies. The low water levels directly lead to a loss of suitable habitat for fish, and the reduced volume of water means that there is less turbulence and thus less dissolved oxygen, indirectly threatening fish life.

The extremes in water flows can be inferred from the characteristics of the stream channel. Typical measurements of an urban stream include the wetted channel width (width of water present) and the much wider bankfull channel width (the width of the channel up the bank that is devoid of permanent vegetation because it is scoured by heavier water flows at certain times of the year). Corresponding to these widths are the wetted depth and the bankfull depth.

Water Quality

Biodiversity can be threatened by local storm drain systems that use the creeks and other surface waters in the city. Chemical pollutants can enter from street runoff and through storm drains. The potential for such chemical pollution is high when industry occurs in the watershed. For example, a sudsing agent accidentally introduced into the storm drain system of Byrne Creek Ravine, in

Burnaby, BC, caused a metre of foam to form on the creek (Schaefer et al. 1992). There were also many instances when the creek was filled with silt due to construction in the watershed. Any salmon or trout eggs in the shallows would have been quickly smothered.

Runoff from streets is also subject to thermal pollution. The blacktop can warm the receiving waters, reducing water quality for some organisms. The warmer water holds less dissolved oxygen and can make the water uninhabitable to oxygen loving fish, such as salmonids.

The amount of dissolved oxygen available to aquatic organisms affects their numbers and diversity. A healthy stream is saturated with oxygen (90–110%) for most of the year. Slow flowing or shallow water will heat up. Organic wastes such as fertilizers and agricultural runoff entering surface waters will consume oxygen directly or indirectly by encouraging algal blooms that consume oxygen when they decompose (Streamkeepers 1996).

Aquatic organisms are also sensitive to acidity. Most prefer a range of pH 6.0–8.5. Outside this range there are usually not enough organisms to maintain communities. The pH of a stream usually depends on the properties of the surrounding soil and bedrock of the draining water. Soils with a high mineral content are usually alkaline. Streams in coniferous forests are affected by the high acidity of the needles from the trees. Rainwater itself is slightly acidic, with a pH of 5.0–5.6.

The cloudiness or turbidity of water is also a factor. Suspended particles in the form of sediment, microscopic organisms and pollutants, restrict light penetration through the water column, affecting algal growth. Sediment can also settle on the bottom and smother eggs and aquatic insects. Turbidity can cause severe problems in areas of urban development and agriculture and sites that have been logged. The problem is greater after a period of heavy rain.

Climate

Large cities are usually warmer than surrounding rural areas. The effect is known as the urban "heat island." It tends to be most dramatic on a clear night.

The urban heat island is a phenomenon where the temperature in the core of a city is an average of 3–5°C higher than in surrounding countryside (Akbari et al. 1989). The heat island is a region of air warmer than that outside the city that has its own distinct flow pattern that tends to keep heat within the city. It is caused by the large amount of heat absorbing surfaces (buildings and asphalt) in combination with the high-energy use found in cities (Bolund and Hunhammar 1999). This heat island effect has also increased the probability of smog episodes (Akbari et al. 1989). Warmer temperatures in cities can increase carbon dioxide emissions from fossil fuel power plants (because of air conditioners), municipal water demand, unhealthy ozone levels, human discomfort and disease (McPherson and Rowntree 1993).

The heat island can extend the growing season of plants, the actual times being determined by the type of plant and degree of the heat island effect. In the heart of London, England, the growing season can be extended by 10 weeks (Gilbert 1991). Thermophilic species requiring warm temperatures such as the tree of heaven (*Ailanthus altissima*) can grow in cities in areas where the temperature may otherwise be too cold.

Cities also change local conditions in wind, humidity and precipitation. In general, the frictional drag of buildings reduces wind speed. However, if there are tall buildings, such as the downtown core of larger cities, the reverse may be true, where the buildings cause the wind to accelerate at ground level. The effect of a tall building downtown can cause a tree at its base to be shaped by the accelerated wind and to have smaller leaves on the windward side of the canopy (Gilbert 1991).

Air pollution can contribute condensation nuclei to the atmosphere over the city that will lead to increased cloud cover and precipitation. It is estimated that cloud cover is 5–10% greater than normal over a large city. Although this leads to slightly more rain, the increased runoff from roads and drainage of building sites more than cancels this out, and the flora of cities instead indicates that conditions are in fact dryer than surrounding areas (Gilbert 1991).

Storm drains empty into local streams, causing floods and droughts that devastate flourishing freshwater ecosystems.

Air Pollution

In most cases air pollution comes from transportation. In Vancouver, British Columbia, it is estimated that about 80% of the air pollution is due to automobiles, 67% from cars. Catalytic converters on the exhaust systems of cars reduce emissions of some pollutants such as carbon monoxide and nitrous oxides. Other common sources of air pollution include petroleum refining, smelting, cement production, wood burning and gas heating.

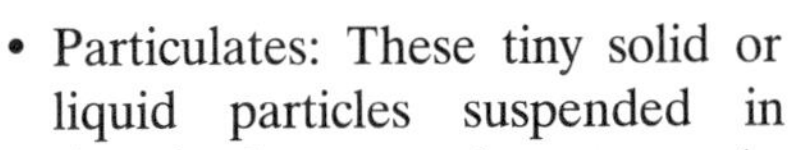

Major air pollutants include:

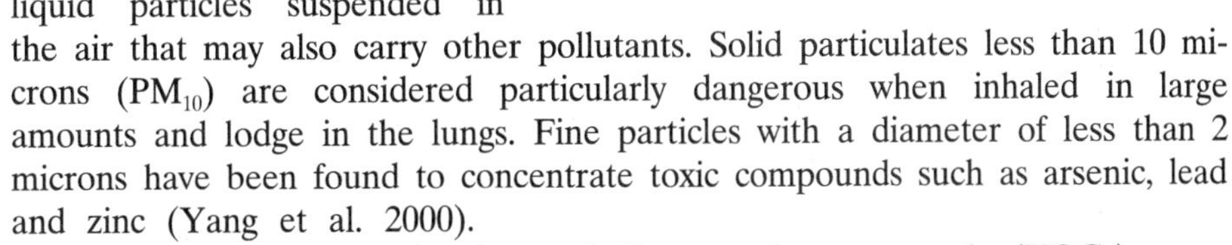

- Particulates: These tiny solid or liquid particles suspended in the air that may also carry other pollutants. Solid particulates less than 10 microns (PM_{10}) are considered particularly dangerous when inhaled in large amounts and lodge in the lungs. Fine particles with a diameter of less than 2 microns have been found to concentrate toxic compounds such as arsenic, lead and zinc (Yang et al. 2000).
- Hydrocarbons: These and other volatile organic compounds (VOCs) come from sources such as gasoline, paint and cleaning solvents. Many of these are

carcinogenic and may also contribute to the formation of other pollutants found in photochemical smog. In addition to being released through combustion, small amounts are regularly released at the gas pump while fuelling.

- Carbon monoxide: This highly poisonous gas comes from automobile exhaust. It combines with the hemoglobin molecule of red blood cells, reducing their oxygen carrying capacity. This can be a concern for people with respiratory and heart problems.
- Nitrogen oxides: These contribute directly to the formation of nitric acid that forms acid rain and indirectly, in the presence of hydrocarbons, to the formation of ground level ozone.
- Ozone: Normally, in the presence of sunlight, the reactions occur as follows:

$$NOx \longrightarrow NO + O$$

$$O + O_2 \longrightarrow 0_3 \text{ (ozone)}$$

$$NO + O_3 \longrightarrow NO_2 + O_2$$

In the presence of hydrocarbons:

$$NO + \text{hydrocarbons} \longrightarrow \text{peroxyacetyl nitrates (PANs)}$$

With less NO (nitrogen monoxide) available to convert ozone to nitrogen dioxide and oxygen, ground level ozone accumulates. The ozone is highly reactive and damages the delicate respiratory tissues of animal lungs and the photosynthetic tissues of plants.

- Sulphur dioxides: These react with water to produce sulphuric acid. As with the nitric acid produced from nitrogen oxides and water, the acidity dissolves the protective waxy cuticles of leaves, making them more vulnerable to infection by microbes and to insect attack. The acidity also leaches nutrients from the soil, creating nutrient deficiencies for the plant.
- Heavy metals: These include lead, mercury, cadmium, nickel and zinc. Many of these are pollutants found in urban runoff originating from tires, brake linings and, historically in the case of lead, gasoline before it was removed as an additive. Metal working industries and shipyards are also major sources.

Factors contributing to the level of the pollutants include the amount contributed from various inputs, the amount of space into which pollutants are dispersed and diluted, and the effectiveness of mechanisms that remove pollutants from the air. Smaller cities in valleys or on coastlines next to mountains may have poorer air quality than larger, more polluting cities on flatter terrain that have their air pollution blown away by winds.

Natural levels of pollution that come from, for example, volcanoes, fires and dust storms, are kept under control through natural processes. Rain removes particulates and the action of soil micro-organisms can convert sulphur dioxide or carbon monoxide into sulphates or carbon dioxide.

Air pollution has a potentially dramatic effect on lichens and vegetation. The epiphytic lichens (growing on branches and trunks of trees) in particular are sensitive and their absence is considered to be an indicator of poor air quality. They rely entirely on the surrounding air to satisfy their requirements. The foliose lichens (epiphytic lichen that adheres to tree bark at several points) disappear first because of their relatively large surface area, then the fruticose species (epiphytic lichen that attaches to bark at only one place). *Lecanora conizaeoides* is the most tolerant of all lichens to toxic chemicals. Although useful as general indicators of air quality they are not very accurate (Gilbert 1991). They take at least five years to re-establish once pollution has disappeared and it is difficult to determine if the boundaries of their distribution are static or moving. Nevertheless, a drop in the diversity, biomass and structural complexity would indicate a deterioration of air quality.

Bryophytes (mosses and liverworts) also show a range of sensitivity to air pollutants. However, losses in biodiversity are often masked by a lusher growth of the few remaining resistant species, such as *Bryum agenteum* and *Ceratodon purpureus*. Similarly, there are changes in the composition and abundance of fungi. Usually, trees are stressed in urban areas and are more susceptible to rusts and mildews. However, oak mildew is suppressed by sulphur dioxide, resulting in many oak woodlands in cities in England being healthier than in the countryside (Gilbert 1989).

As with other taxonomic groups, air pollution has a number of effects on plants that can be acute or chronic. Acute impacts appear after a few hours or days and result in patches of dead cells. Chronic impacts take much longer, and can lead to a partial destruction of chlorophyll, a reduction in productivity, and a decrease in the width of annual growth rings in trees. The chronic, less severe effects may show themselves as yellowing leaves or variegated leaf colours. Severe impacts lead to stunted or deformed growth and death. Deformed tree growth due to air pollution may appear as dense and compact or flat-topped canopies with twisted branches. If the air pollution subsides, there can be a period of rapid new growth that will produce a "double-canopy" effect (Gilbert 1991).

Foliose lichens such as this old man's beard are indicators of good air quality.

Gaseous pollutants can directly affect the plant's body or indirectly affect the plant through the soil. Deposits of phytotoxic residues such as lead, copper, tin and arsenic destroy plant tissue on impact. Air pollutants also generally act as condensation nuclei in the atmosphere, refracting light and affecting the light spectrum available to the plant for photosynthesis. Finally, dust particles settle on leaves and block light from the chloroplasts, thereby reducing photosynthesis and plant productivity.

Air pollution affects plants unequally. Lists of the most pollution resistant conifers and broadleaves are provided below (Lanzara and Pizetti 1977).

Among the most resistant conifers are:

European larch	*Larix decidua*
English yew	*Taxus baccata*
Blue spruce	*Picea pungens*
Omorika spruce	*Picea omorika*
Cedars	*Thuja sp.*
Juniper	*Juniperus sp.*
Dwarf Pine	*Pinus mugo*
Black Pine	*Pinus nigra*
Cembrian pine	*Pinus cembra*

Among the most resistant broadleaves are:

Speckled alder	*Alnus incana*
European birch	*Betula pendula*
Mountain elm	*Ulmus montana*
Sycamore maple	*Acer pseudoplatanus*
Norway maple	*Acer platanoides*
Northern red oak	*Quercus borealis*
Poplars	*Populus sp.*

Animals suffer the same losses to biodiversity due to acute and chronic effects of air pollution. Pollutants affect delicate tissues of lungs, skin, and gills that may be used for gas exchange. Amphibians in particular have a moist, sensitive skin to allow for gas exchange. This same sensitivity makes them more vulnerable to acidic pollutants in the air and they are usually the first species to disappear when there is air and water pollution.

Heavy metals such as copper, lead, nickel, cadmium and mercury, and synthetic organics such as benzene, chlorophenols and pesticides can bioaccumulate in tissues and biomagnify through the food chain. The significance of this is that even small amounts of these substances in the environment can lead to serious problems for the fish and for bird predators higher up in the food chain. Examples include the effect of DDT on the eggs of falcons and dioxins on the eggs of herons; the shells in each case formed unnaturally thin due to these chemicals and

A northern red oak is one of the more pollution resistant species.

break more easily, reducing reproductive success.

These chemicals are often powerful carcinogens as well and can cause cancers in wild animals, not only in humans. They cause skin lesions on fish that live in bottom sediments where these chemicals tend to accumulate in water. The toxins accumulate in the livers of other animals and cause lesions, which eventually destroy the organ. Some fish species, such as salmonids, are intolerant of water pollutants, whereas other species, such as catfish and carp, are more tolerant.

Urban environments are unique, with their heat island, increased carbon monoxide and acidic particles, haze-light spectrum (altered abundance of wavelengths of light compared to natural lights due to turbidity — particles — in the air from pollution), altered drainage patterns of infiltration/runoff and altered soil chemistry due to acidity impacts on nutrients. However, despite their highly artificial composition and pollution problems, urban areas are not biological deserts. The city can be a complex habitat suitable for many species, with its buildings creating unique environments of temperature, moisture, and acidity (Oke 1976). Gardens, yards, parks, and recreation areas establish significant areas of green space, with linkages that can support a surprising array of plant and animal life (Laurie 1979). Species and population diversities can be high, although in many cases they involve introduced species.

In the case of people, a major part of our urban environment is indoors. Indoor air quality can be a concern. Sources include:

- Formaldehyde from particle board furniture and carpets
- Cooking fumes — food burned on the stove or in the oven
- Cleaning agents
- Air fresheners and disinfectants
- Smoking
- Fumes from glues
- Aerosol sprays
- Pesticides
- Radon gas
- Asbestos

Many of the air quality concerns from fumes in homes can be mitigated by proper ventilation. Particle concentrations in urban microenvironments vary significantly between areas with and without combustion sources (Levy et al. 2000). Larger particles are elevated around human activity and smaller particles (PM_{10}), being significantly higher on buses and trolleys than outdoors. Outdoor environments with smokers have slightly elevated particle counts and mass concentrations. Elevated levels also occur in indoor environments with central air conditioning.

Collisions with Structural Hazards

Structural hazards such as tall buildings, smoke stacks and transmission towers in cities may actually kill more birds than oil spills (www.flap.org). The Fatal Light Awareness Program determined that hundreds or even thousands of birds may be killed in one evening when they are attracted to the lights of a skyscraper, blinded by weather and unable to see glass. Perhaps 100 million birds are killed this way in North America each year. In Toronto, Ontario, alone, 130 species are affected. Even if the birds do not die in a collision, they are reluctant to leave the light at night and may fly until they are exhausted and drop to the ground, vulnerable to starvation or predation.

Similarly, some birds may fly into glass during the day. A window may reflect surrounding vegetation or the sky and be invisible to a bird (www.flap.org).

Human Diseases

Pigeons and European Starlings are the main carriers of two fungal diseases, cryptococcosis and histoplasmosis, that affect people. Droppings from pigeons frequently contain the yeast *Cryptococcus neoformans*, which is the dominant organism living in the accumulated alkaline nitrogen-rich debris in pigeon roosts, where it can persist for years and represent a health hazard for people. Pigeons themselves are not affected. *Histoplasma capsulatum* is a fungus found in areas where the guano of birds and bats has enriched the soil. The highest occurrences of the disease are in places with high concentrations of starlings (Gilbert 1991).

Other diseases of concern to people and other wildlife include

- Hantavirus: *Bunyavirus* sp. is contracted through inhalation or contact with mouse saliva, urine and feces.
- Rabies: The *Rhabdovirus* sp. virus can be spread from a bite or inhalation from carnivores and bats.
- Salmonellosis: The bacteria *Salmonella* spp. and *Campylobacter sp.* that is similarly transmitted through ingestion.
- Lyme Disease: A bite from the tick *Ixodes dammini* can carry the bacteria *Borellia burgdorferi.*
- Plague: This disease that historically killed millions came from the bacteria *Yersinia pestis* that was carried from rats to people by fleas.
- Ringworm: This is actually a fungal disease caused by *Microsporum* sp. and *Trichophyton sp.* from direct contact.
- Parasites: A number of parasitic infections are caused by the ingestion of the eggs or cysts of the organisms involved, usually protozoans (single-celled organisms) and nematodes (round worms). These organisms include: *Toxoplasma gondii*, *Echinococcosis* sp., *Baylisascaris* sp. and *Giardia* sp. They are transmitted by birds and mammals and some, like *Giardia*, are of particular concern if they infect drinking water supplies for a city.
- Fecal Coliforms: Bacteria such as *Escherischeri coli* from human sewage or from large concentrations of waterfowl, especially Canada Geese (*Branta canadensis*), can cause infection through ingesting water from affected streams and lakes.

Urban Design and Wildlife Habitat

Some habitats in cities are valuable to most wildlife, especially if the area is large and offers living conditions suitable for interior species (ones intolerant of disturbance by other species present along edges). However, the fragmented nature of most habitats, their distribution in the city and the collective benefits of marginal habitats are just a few of the factors contributing to urban biodiversity. What follows is a discussion of some of the attributes of urban habitat that determine its value in supporting biodiversity.

Patch Size and Shape

Urban development leads to habitat changes, such as decreased patch size (a patch is a habitat fragment), further distances between patches and more edge habitat at the expense of interior habitat (Hansen et al. 1995, Reed et al. 1996). The condition of the habitat in the patches may also change where a patch may remain large, but the habitat within the patch may deteriorate because of urban development. The smaller patches with greater edge are vulnerable to disturbance by introduced species, such as black locust (*Robinia pseudoacacia*) in Seoul, Korea, that regularly invades patches (Hong et al. 2003). Patch dynamics include more than just their spatial pattern at any given time. How these patterns change over time and the effects of this change on ecological and social processes are also important (Grimm et al. 2000).

Smaller animals may be able to use small patches of habitat, but larger mammals will need larger areas. Although smaller patches may not be able to sustain breeding populations of species, they are valuable in providing feeding areas and refuges. Larger patches have more potential habitats that can sustain populations of wildlife. In a study of urban woodlands affecting breeding bird diversity and abundance, Tilghman (1987), determined that it is the "size and vegetation diversity and interconnectedness of patches which determine population size and animals using the habitat". He found that areas less than 101 hectares could only support wildlife with narrow habitat requirements, such as smaller mammals. Larger mammals, such as coyotes (*Canis latrans*), bobcats (*Lynx rufus*) and hawks, needed interconnected patch areas that were between 101 and 4,900 hectares.

A study of a cemetery that was over 40 hectares in size and had many mature trees and several water features had about "42 common or resident birds," as opposed to a residential yard one tenth of a hectare in size that only had about "12–15 resident birds species" (Tilghman 1987). Larger habitats are more favorable for more species of birds.

Kennedy et al. (2003) suggest in their survey that habitat patches of 55 hectares (137.5 ac) for birds capture 75% of species requirements. They indicate that such a minimum land parcel will not, however, provide for area-senstive species such as wide-ranging predators or interior bird species. The latter may require areas greater than 2,500 hectares (6,175 ac). Small mammals, such as rodents and rabbits, have a minimum patch size of 1–10 hectares (2.5–25 ac). Bears have a recommended patch size of 900–2,800 hectares (2,223–6,916 ac), and cougars 220,000 hectares (543,400 ac). One study about fish found that a suitable patch size for a watershed was 2,500 hectares (6,175 ac) for bull trout (*Salvelinus malma*). An overall rule of thumb would be for land use planners to conserve 20–60% of the natural habitat in a landscape to conserve biodiversity. Invertebrates and plants have smaller dispersal ranges, and may need habitat areas of less than 10 hectares (25 ac).

The shape of the patches also influences the biodiversity a habitat can support. Edge effects become important for determining the quality of species for biodiversity indices. Edge effects often occur when a natural climax community is disturbed and habitats become fragmented. For example in a forest, generally there is a large tree canopy and very little light can reach the forest floor; most species present are tall trees. When trees are cut, more light can reach the forest floor allowing more species of vegetation to grow and hence more species of wildlife. In urban areas, edge is created when vegetation is planted in narrow patches or there is a sharp division between mowed grass and tall grassy margin. In both these instances, cover is compromised along the edges of the vegetation feature. Often nesting success is decreased due to the increase in predation because there is less space for cover.

With the variability in the habitat that has more light and less cover, more species are able to use edge habitats.

The greatest diversity of vegetation offers up the most to wildlife and, therefore, the diversity of wildlife will similarly increase. However, edges can also cause an increase in nest predation by crows, as well as nest parasitism from Brown-headed Cowbirds (*Molothrus ater*) (Yahner 1988). Therefore species numbers and abundance are not necessarily the best indicators of habitat value. Rather, it is the quality of the habitat and the quality of the species that use the habitat that are important for determining the value of an urban habitat. If the species present are mainly generalists, the habitat may not be as valuable for all ranges of wildlife. For example, it is possible to see crows, gulls and starlings on city streets. These species are generalists, and will feed on whatever they can find. Species that have specific habitat and diet requirements, such as a woodpecker, for example, would not survive on a city street.

Patch Size Decay and the Species Area Curve

As is the case with many causal relationships in ecology, the reduction in biodiversity that occurs as a result of habitat destruction in cities is not linear. Patch size decay is a functional response. A reduction in 10% of the habitat does not necessarily reduce biodiversity by 10%. Destroying one hectares of habitat in the wilderness fringe at the edge of a city will probably have no impact. However, once a significant amount of habitat is already lost and natural ecosystems are at a critical point, even small losses of habitat in the remaining refuges can have dramatic effects on biodiversity. Alternatively, in areas that are already degraded with little biodiversity, even extensive development of housing or commercial sites may have no impact on biodiversity.

What is of interest is the marginal habitat losses in assessing their impacts on biodiversity. In a plot of habitat destroyed against largest patch size, there is a curve, not a steadily decreasing straight line. The plot curves gently downward for a while, where reductions in patch size result from comparable levels of habitat destruction. However, there is a threshold where dramatic reductions in patch size produce little habitat loss. The patch size has crossed a threshold and the consequences of habitat loss are severe beyond this point (Sole and Goodwin 2000).

The pattern of patch size decay is related to the species area curve in ecology. In this plot of numbers of species against area of habitat there is, again, a functional response. As the area begins to increase there are comparable increases in the numbers of species. However, beyond a certain point, even small increases in area result in large increases in the numbers of spaces. The rate of species addition with area increases steadily, crosses an inflection point where the species still increase but not as quickly, and then eventually levels off at a point where increasing the area further adds few species.

In addition to the size of a patch, the quality of the habitat within a patch can change and produce a similar response. In cities, it is clear that large remnant forests protected in parks are excellent habitat supporting a great deal of biodiversity. At the other end of the spectrum we have downtown or high density residential development with little biodiversity. In between is a spectrum of patch sizes and habitat quality. An important question is, where is the threshold? What habitats are on the steep slope of the species area curve or a curve of patch size decay? Are they smaller remnant native ecosystems? Or can they be golf courses and cemeteries?

Measures of Patch Characteristics

The area of a patch is but one measure of its value as wildlife habitat. Other patch characteristics include (after Forman and Gordon 1986 in Morrison et al. 1992):

- Shape of Patch
- Isolation of Patch
- Accessibility of Patch
- Interaction Among Patches
- Dispersion of Patches

A shape index can be calculated to assess edge effects by the formula:

$$D_1 = P / 2(A\pi)^{1/2}$$

In this formula, D*i* is the shape index for patch I, P is the perimeter of the patch and A is the area of the patch. The calculation is intended to show the relationship between the perimeter and area as a measure of the width or narrowness of the patch and its relative roundness, all of which, in turni reflect the degree of edge effect. Similar calculations from fragmentation models can also be performed for the other patch characteristics.

Concerning the above characteristics, a crude measure of isolation is to look at the average minimum distances between patches, in other words, a measurement of aggregation. If the patches are less isolated there will be more interactions between them, although the effect of isolation largely depends on the species involved. A larger mobile mammal, such as a raccoon, can tolerate more isolation than a small, less mobile reptile, like a turtle.

Any one of these characteristics, or several in combination, can play a major role in determining the distribution and abundance of certain species in urban areas.

Quality of Habitat

Quality of habitat is very important for wildlife. The quality of a habitat can be measured by looking at the quality of the wildlife species using the area. "Forest interior" species, such as the Warbling Vireo (*Vireo gilvus*), that are more affected by changes in habitat, are good indicators of habitat quality. The presence of more sensitive species would reflect the habitat quality and these quality habitats should be managed to try to preserve their integrity.

The original wildlife diversity and structure for an ecoregion determine the quality of habitat and biodiversity that could exist in an urban area (http://www.on.ec.gc.ca/wildlife/docs/forest.html). Having the knowledge of the types of species that should be present in the area if the forest were untouched is an index for determining the health of an ecosystem.

Although a pristine forest or grassland ecosystem is not a reasonable goal, it is possible to add features to urbanized areas that increase the habitat value to allow for a greater diversity of wildlife. Such a feature would be a pond with

Larger mobile predators such as this GreatHorned Owl may benefit from having many smaller habitat nodes closer together.

native vegetation along the banks that would provide food, cover and resting areas for many species of birds.

Water Features

The presence of water increases the numbers and types of species. A study of golf courses revealed that water bodies attract the highest density of birds over other types of habitat (Moul and Elliot 1994). The other habitats that are used in order of preference are hedgerows, trees and then turf. (Moul and Elliot 1994).

Location

The location of a habitat with regard to its surrounding environment is important for determining its value in supporting biodiversity. A golf course or cemetery would have a greater impact on native species of wildlife or forest interior species of wildlife if they are in close proximity to natural "pristine" forest at the city's edge. Conversely, areas bounded by busy thoroughfares would not be able to serve the same types of wildlife due to the types of habitat that surround them.

Connectivity

Another factor to take into consideration when dealing with habitat values of areas for wildlife is their connectivity with other green spaces. Connections are needed to provide migration routes for various species of wildlife between larger nodes of green space. It is sometimes more useful to consider a population within a green space in the larger context of the metapopulation to which it belongs and that encompasses other green spaces nearby. The size of corridor or connection and the types of vegetation present in these areas, as well as the type and size of green space node, determine the types of wildlife that can use an area. For example, it would not make sense to plant forage for deer in a highly urbanized area with large forested area nearby where the deer could live.

In urban environments there is usually one large green space, or mother node, in a metapopulation zone that has significant influence on the surrounding area. As the demand for land to develop grows with the population, cities can usually only afford to preserve a few large green spaces. These green spaces tend to have high biodiversity and provide important breeding and seeding habitat for interior species, as well as edge species and transients. Smaller green spaces or satellite nodes may not be able to support large numbers of species on their own, but are able to provide important peripheral habitat to species in the mother node. Satellite nodes are partly or entirely dependent on individuals immigrating from the mother node (Hansson 1991). They have a higher rate of extinction than the mother nodes and, therefore, need to be repopulated constantly (van Apeldoorn et al. 1992). This requires proximity to the mother node. As the urban environment becomes increasingly fragmented, satellite nodes get smaller and further away from the mother node, making dis-

Figure 2.1 Map of a connectivity analysis study area in Coquitlam, British Columbia.

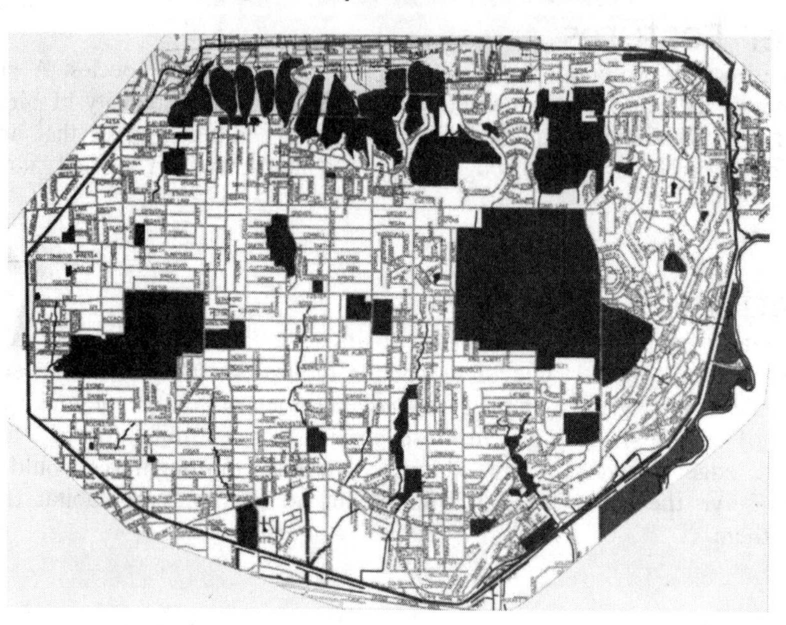

Green spaces are shaded in black and the lines are streets. The metapopulation zone is an area bounded (dark line) by Lougheed Highway, North Road, Clark Road, St. John's Street, and Barnet Highway. The total area of the zone is approximately 2,000 ha. The mother node, Mundy Park (largest green space), is about 175 ha. Map not to scale. (Rudd et al. 2002)

persal even more difficult. As a way of preserving the biological integrity of a landscape, corridors and habitat matrices must be in place to allow dispersal between green spaces.

Several potential networks can be generated with greenways to join satellite nodes and a mother node. There are two major groups of network models, branching and circuit. An example of a branching network is the Paul Revere model (Linehan et al. 1995), one of the simplest network models connecting all nodes. It is also the cheapest to create for the group concerned with creating the network.

The other family of networks is circuit networks. These networks tend to be more complex than branching networks and often represent a lower cost to the user: the flora and fauna using the green spaces as their habitat and benefiting from the networks. Examples include the Travelling Salesman and the Least Cost to User (Linehan et al. 1995). The Travelling Salesman is the simplest, where each node is connected only to two other nodes. The Least Cost to User is the most complex network model because all nodes are directly connected to each other.

The Paul Revere model, a simple network with only one connection of one node to another, would probably be used by migrating songbirds as they move through a city. Circuit networks tend to be more complex than branching networks and often represent a lower cost to the user; the flora and fauna using the green spaces as their habitat and benefiting from the networks. The Traveling Salesman (a simple loop), is probably applicable to resident animals or breeding birds with particular habitat needs, and the Least Cost to User, available to animals that are not shy to travel through backyard habitat.

The importance and significance of these networks can be evaluated using Gamma, Beta, and Cost Ratio indices. The Gamma Ratio represents the percentage of connectedness within each network. It can be determined by dividing the number of links in the network by the maximum number of possible links. The Gamma index ranges from zero to one, and the closer to one, the greater the degree of connectedness within the network. The Beta index indicates the complexity of the network. It is calculated by dividing the number of links by the number of nodes. When the results are less than one, the network is open or branching. If the result is one, the network is a single circuit, and if Beta is greater than one, there is greater complexity within the network. The Cost Ratio index indicates the relative cost to both the user and the builder. It is calculated by subtracting one from the product of the number of links in the network by the total distance of those links. The closer to one the Cost Ratio is the greater cost to the builder and the lower cost to the user.

Dispersal between nodes, which is simpler in a well-connected network, is essential to prevent inbreeding depression and the disease and extinction that follow. However, as the number of links decrease the ease of dispersal also decreases. This increases the probability of extinction (Noss 1983, Schippers et al. 1996). The major goal of connectivity is to protect the integrity, structure, and function of ecosystems.

The relative ecological value of connectivity can be estimated by considering the area of a habitat node and its distance to other nodes. Generally, areas have a greater interaction when they are larger and closer together (Linehan et al. 1995). Connectivity using the gravity model that introduces a weighting factor for a node is determined as follows:

$$Gab = (Na \times Nb) \div Dab^2$$

Gab — level of interaction between nodes a and b.
Na — nodal weight of node a.
Nb — nodal weight of node b.
Dab — distance between nodes a and b.

The weight is the habitat size required for the particular species being considered, perhaps an indicator species of overall biodiversity. With smaller nodes it will be important for weightings to consider effect distances, influenced by traffic volumes and speeds, from surrounding roads. An apparently suitable node may, in fact, have no habitat for interior species in which case it should be considered entirely as edge habitat or part of the diffuse texture. It may also be difficult and inadvisable to control for invasives in places where vehicle traffic plays a significant role in their dispersal. The gravity model provides an unbiased method to determine different levels of interactions between nodes.

Creating corridors using the connectivity analysis is much more effective than randomly selecting links. Randomly selected networks may not be as effective at protecting and enhancing biodiversity. Both the theory of island biogeography and metapopulation dynamics assume that suitable patches of habitat are interspersed with uninhabitable areas (Andren 1994). This creates a divided landscape. Therefore, it is important to remember that preserving parks is only part of the solution. Without connections between them, isolation and loss of genetic diversity is imminent (Hobbs & Saunders 1990). Green corridors, utility rights-of-way, and backyard habitat are important parts of urban planning, as they increase biodiversity in cities and improve the quality of life for all residents. For example, they increase opportunities for wildlife viewing, human relaxation and education, as well as controlling pollution, temperature and climate, erosion, and noise (Adams and Dove 1989).

Taylor et al. (1993) defines landscape connectivity "as the degree to which the landscape facilitates or impedes movement among resource patches," whereas connectedness refers to structural or physical connections between patches or nodes. Where resources are available, future studies should concentrate on connectivity rather than connectedness. However, this poses another problem as examinations of connectivity are usually conducted on a single species (e.g., Bennett et al. 1994) and the results may not be transferable to all species in an area. Analyzing structural connectivity may present more general results.

It is difficult to recommend a particular corridor width because the effectiveness of a corridor also depends on its length, the continuity and quality of the habitat and the topography. A corridor width of 32 m has been found to encourage the movement of butterflies and 100 m to encourage the movement of birds (Kennedy et al. 2003).

Several guidelines have been created for corridors in various habitats. Environment Canada's Ontario division created some Forest Habitat Guidelines and these guidelines outline the sizes of patches and corridor width sizes. These were based on independent study as well as research on previous studies performed on various types of wildlife and their habitat requirements. Based on the requirements of various species of wildlife, a minimum width of 100 m should be available for movement of species with a minimum width of 500 m for more specialist/sensitive species (http://www.on.ec.gc.ca/wildlife/docs/forest.html). Patches of forest present should be square or circular in shape at least 200 hectares in size and a minimum of 500 m in width and be within 2 km of another patch. (www.on.ec.gc.ca/wildlife/docs/forest.html).

Therefore, the greater the forest area, the greater the number of species it will support. But the effect of forest patch size is dependent on other factors for different species of birds in different geographic areas. "Factors such as forest shape, proximity to adjacent forests and presence of specific forest microhabitats combine with forest size to give birds a better chance of avoiding excessive predation and parasitism and achieving breeding success." (www.on.ec.gc.ca/wildlife/docs/forest.html).

Green Cities, Eco-cities and Environmental Stewardship

There have been essentially three different visions of city design that directly or indirectly benefit urban biodiversity.

Green Cities

At the turn of the century there was a garden cities movement that concentrated on view corridors from homes and vantage points in the city. It was a movement to create colourful gardens that were pleasing to the eye. The green cities movement is a modern interpretation of the garden city, broadening the concept to harmonize urban design with natural processes. Michael Hough (1984) provided a blueprint for a city that took advantage of natural processes to provide some of its functions and promoted the natural environment as a way to improve livability.

Eco-cities

A more recent movement was founded by Richard Register (1975). He presented a plan for a future Berkeley, California that was in harmony with the bioregion constructed on the principles of life, beauty, and equity. His was a vision of a three dimensional city with a compact design incorporating reduced street widths and high rise, high density housing forming integral neighbourhoods that provided access by proximity rather than transportation. Roseland (1997) elaborates on the dimensions of the eco-city that include:

- Healthy communities providing peace, food, shelter, education, and other necessities for a healthy and happy population
- Appropriate technology suited for the local setting (e.g., solar energy)
- Community economic development where communities assume responsibility for their local economy
- Social ecology promoting social equality and justice
- The green movement (ecology, social responsibility, grassroots democracy, nonviolence)
- Bioregionalism where economic and political structures are decentralized and redefined along natural boundaries (such as a river basin or watershed)
- The First Nations world view in which nature is seen as an extension of the family and forms a society of beings
- Sustainable development whose tenet is that today's generation should not compromise the survival of future generations

The eco-city is an ideal. There is no city that embodies all the dimensions outlined by Roseland and it is unlikely that there ever will be one because of planning time frames (need long-term vision and commitment), politics and social and economic crises. Nevertheless, the eco-city is a valuable concept, a goal to work towards. Although the eco-city seems a remote reality for an entire city, many of its dimensions may be possible to approximate at a smaller community or neighbourhood level within a city.

Environmental Stewardship

There was a renewed interest in wildlife habitat in cities in the 1980s that manifested itself as environmental stewardship. A number of events may have contributed to this focus on wildlife in the city, unattached to the human desire to see pretty landscapes or embedded in a larger framework that included human social and economic issues. The design of the city basically stayed the same, but citizens took it upon themselves to look after animal habitat.

Environmental stewardship followed a campaign in many regions by nongovernmental organizations and, later, government promoting environmental citizenship, initially focused on reducing energy consumption and solid waste, but later broadened to include the environment as a whole, including biodiversity. The public was mobilized and wanted to do its part for the environment.

There had been some success in having a number of areas set aside as protected areas, with the arbitrary goal of 12% of the land base established by the Brundtland Commission in 1987. Nongovernmental organizations observed that the protected areas were often mountaintops and otherwise biologically unproductive areas. In subsequently identifying what areas might be more important, cities were found to occur in biologically very productive areas such as floodplains and estuaries.

The connectivity value of a corridor increases as the biodiversity in the corridor more closely approximates that of the habitat nodes it connects. Planting native shrubs in utility rights-of-way improves their contribution to connectivity.

For whatever reason, the wildlife value of natural habitat in urban areas, in and of itself, is the subject of much interest and effort. There is considerable public support to protect the remaining natural areas in cities as parks. The primary areas of focus for restoration include streams and backyards. There are numerous stream stewardship groups and several backyard habitat programs (Naturescape BC by Pincott and Campbell (1995) is one of the best).

Perhaps most encouraging are attempts to create a biodiversity conservation strategy for cities. This is most evident in discussions on connectivity where islands of habitat are connected to create larger functioning units supporting better gene flow, larger food webs and larger breeding populations. The Ecology of Greenways (Smith and Hellmund 1993) is an excellent example of this approach. Although the term greenway is new, emerging in the 1950s, the potential of linear open spaces as part of a park system in a city was recognized in the 1860s by Frederick Law Olmsted, the designer of Central Park in New York.

As the number of people living in cities continues to increase, the amount of land required for residential and commercial development will also increase. This inherently causes urban biodiversity to decrease since undeveloped land is usually much more attractive and required as building sites, not green space. As explained above, there are movements to create more green space and wildlife habitat in urban areas. However, when potential land use is in question, development pressures are often a priority over leaving an area as green space. Natural systems and ecosystem functioning, both within and outside the city, supply many benefits that support and ensure human survival.

Marking storm drains in the community with yellow fish painted on the pavement is one approach to environmental stewardship, drawing attention to the connection between the street and fish habitat.

3

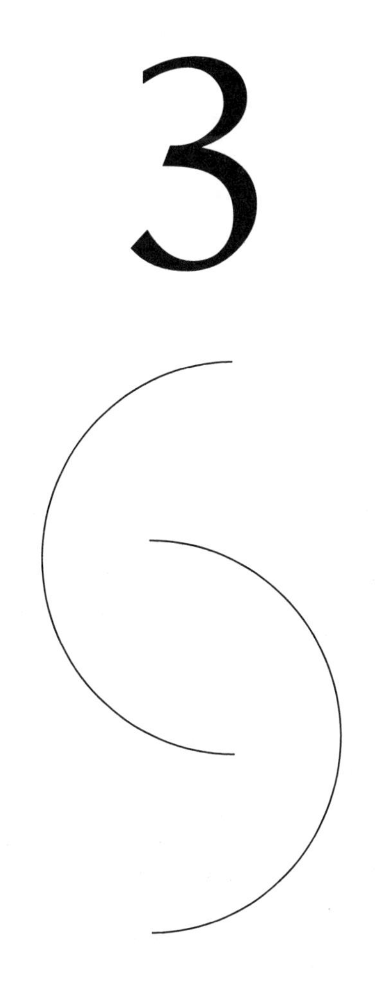

Indicators of Urban Biodiversity

LOSS of natural habitat and invasive exotic species cause a loss of ecosystem diversity, and result in extensive local extinctions of native species (see Thomas 1984; Heath 1981). The expansion of cities into the natural environments along their periphery is a major factor contributing to global losses in biodiversity. Modern urban landscapes with their extensive blacktop and lawns only became common after World War II due to the prevalence of the automobile and rapid population growth. Urban impacts on the natural landscape are great. For example, about 2% of the entire land base of the United States is composed of lawns (Garber 1987).

Murphy (1988) points out that urban areas are synonymous with a loss of biodiversity. A number of forces are at work besides the direct removal of natural habitats. Reductions in biodiversity can also occur from air and water borne pollutants, the overdrafting of local aquifers, and changes in groundwater flows caused by storm drains. Pets and livestock may prey on native animals, and spread diseases and parasites among more vulnerable native species.

Despite their highly artificial composition, urban areas are not biological deserts. The city can be a complex habitat suitable for many species, with its buildings creating unique environments of temperature, moisture, and acidity (Oke 1976). Gardens, yards, parks, and recreation areas create significant green space with linkages that can support a surprising array of plant and animal life (Laurie 1979). Species and population diversities can be high, although in many cases they involve introduced species, and they rarely form fully functioning ecosystems.

Better natural habitat can be found on the steep slopes of ravines and canyons spared from development. These areas can act as refugia for the original native vegetation of an area. Well known examples of such areas are Rock Creek Park in Washington, D.C., and Fairmont Park in Philadelphia (Murphy 1988). The ravines and canyons represent hidden natural assets by providing relatively high biodiversity in the heart of residential development. The ravines of Toronto, Ontario, are described as the city's best kept secret. The Humber, Don and Rouge Rivers, along with Etobicoke, Mimico and Highland Creeks, offer excellent wildlife habitat and trails to explore (Seymour 2000).

It is difficult to measure biodiversity and no single descriptive statistic can accurately assess the complexity of the ecosystems, species, and gene pool characteristics that contribute to biodiversity (Pielou 1973). In this book habitat quality and biodiversity are assumed to be directly associated, and both can be assessed by using indicator species. The suitability of this habitat for other associated species will help assess the potential level of biodiversity. Indicator species are best selected based on a number of criteria. No indicator can possess all of these properties. Therefore, a set of complementary indicator species can be used to assess the status of biodiversity over time.

Figure 3.1 Criteria for Selecting Indicator Species

1. **Sensitive to environmental stresses.** The first signs of environmental stress usually occur at the population level, affecting sensitive species (Welsh & Droege 2001). An effective indicator species should then be sensitive to any physical, chemical or biological stress, so as to provide an early signal reflecting the health of its ecosystem.
2. **Specialist.** In terms of resource or habitats used, each species falls somewhere along a specialist to generalist continuum. Specialists are more sensitive to habitat changes and meeting their needs is assumed to provide the needs for generalists as well. However, managing for specialists may not provide suitable habitat conditions for all species. Therefore, indicators were not selected solely on this criterion, but on how strongly the indicator species is associated with the habitat attributes of interest.
3. **Permanent residents.** Migratory species are subjected to a variety of sources of mortality and a measured decline in their abundance may not be related to habitat conditions on the breeding grounds.
4. **Regional breeders.** This criterion applies mainly to birds and also fish species that are anadromous, since some migratory species are indicators. A healthy breeding habitat is sometimes critical to sustain their population levels.
5. **Native.** Native species are those that occur in a region in which they evolved. In North America, they include species that were present prior to European settlement. They make effective indicator species because they have many ecological advantages to contribute and help maintain biodiversity. For example, native plants are well adapted to local environmental conditions. They can maintain or improve soil fertility, reduce erosion, provide familiar habitat for local wildlife and contribute to the overall health of natural communities.
6. **Habitat requirements similar to those of other species.** Indicator species whose life cycles are intertwined or correlated with other assemblages (lesser-known) are valuable to monitor. A suitable habitat for one species may indicate that the habitat is suitable for other species that are more difficult or expensive to monitor.
7. **Habitat heterogeneity.** Habitat diversity plays a prominent role in the determination of species richness and there is often a positive correlation between habitat diversity and species diversity (Hermy & Cornellis 2000). Therefore, indicator species should represent and reflect various habitat types.
8. **Visible and identifiable.** Indicator species should have some appeal with the general public. To protect biodiversity, public awareness and understanding are necessary to support conservation initiatives through expenditure of tax dollars. Easy to recognize species are more identifiable to the public and the benefits or information about them with carry more impact.
9. **Easy and cost-effective to monitor.** Need to be able to measure, collect or assay indicator species long term to assess the status of biodiversity over time.

Species Diversity Along the Rural — Urban Gradient

Due to the ecological and physical differences between urban and rural environments, species composition and diversity vary along the rural-urban gradient (Pickett et al., 2001; Blair, 1996). Habitat specificity and adaptability determine which wildlife and plant species are able to tolerate or be displaced by urban development and habitat fragmentation. This is influenced by the relative costs and benefits associated with obtaining resources and avoiding predation (Rubin et al. 2002).

Generally, moderate levels of urban development may increase the number of species present, but this is largely due to the increased presence of exotics (Pickett et al. 2001). This was found in a study by Blair (1996), who observed an increase in overall species diversity in birds, but a decrease in native bird diversity. Even with regard to vegetation, a pattern exists where the number of native species decreases from the urban fringes to the centre of a city (Pickett et al. 2001).

As the landscape becomes increasingly dissected into small, isolated fragments, the level of biodiversity can decrease as "edge" species become more prominent (With & Crist, 1995). In many cities, the shift towards greater development in many suburban areas is clear. Care must be taken so that the diversity within urban areas does not diminish to include only exotic species or "edge" species capable of adapting to and exploiting disturbed habitat. While some wildlife and native plant species avoid urban areas altogether, others continue to use and thrive on developed areas as their preferred habitat. Many wildlife and plant species are able to adapt to new environments given favourable conditions and optimal forage opportunities. Generally, urban species can be divided into three categories: 1) urban avoiders, 2) urban adaptors and 3) urban exploiters.

Urban Avoiders

Species termed "urban avoiders" are particularly sensitive to human-induced changes to the landscape (Blair 1996). They typically avoid cities and are found in greatest abundance and density in natural, undisturbed habitats. For these species, urban development is the equivalent of habitat loss (Rubin et al. 2002), which cannot be replaced or substituted for easily. Examples of urban avoiders include the tailed frog and Douglas squirrel.

Urban Adaptors

Urban adaptors include species that are commonly found in rural or more natural environments, but these species have learned to exploit additional resources, such as ornamental vegetation found in moderately developed areas (Blair 1996). These species have adapted to urbanized sites and have been successful in locating new shelter and food resources and have avoided predation and disease. However, urban adaptors still prefer natural, undisturbed habitats.

Indicator species included in this study that have expanded their habitat and have adapted to the changes that have come with urban development include the Northern Harrier, Pileated Woodpecker and big brown bat.

Urban Exploiters

Urban exploiters include species that actually prefer disturbed habitat caused by urban development. Their densities are higher in urban environments, since they are able to increase their fitness by exploiting new food sources (Rubin et al., 2002) in the absence of natural predators. Urban exploiters are commonly introduced species that are generalists. Examples of urban exploiters include European Starlings, House sparrows, eastern gray squirrels and feral or free roaming domesticated cats.

Indicator Species of Disturbed Ecosystems

Indicator species of disturbed habitat are typically generalists, non-native and invasive. The indicator species of disturbed habitat for this report includes those that can either contribute to the loss of species diversity (invasive species) or those that occupy habitat of depleted biodiversity (urban exploiters). Some indicator species of disturbed habitat may fall into both categories. Due to their high tolerance of urban development and their ability to exploit several ecological niches and out-compete native species, they are considered to have a negative impact on biodiversity.

Canada Geese are a common sight in cities and are a good example of an urban adaptor.

Non-native species pose a serious threat to biodiversity because of the absence of natural controls, such as insect pests and competitors. This allows them to grow and establish themselves in new areas, where they can successfully out-compete native flora or fauna. In the case of non-native vegetation, they can significantly degrade entire plant communities and disrupt ecological processes. Common native plants can be crowded out, and wildlife that depend on their existence will either have to adapt or be displaced. Endangered species have in some cases been driven from their last habitat by invasive non-native plant species.

Urban exploiters are capable of increasing their fitness and numbers by taking advantage of the food and shelter resources available within an urban environment. Disturbed areas that are open and fragmented, such as roadsides and vacant lots, seem to be the preferred habitat for some non-native species. In these cases, the introduced species may not physically be the cause of displacement for native species, but do well in a habitat that has been disturbed. This makes them indicative of a habitat that has been disturbed, degraded or unsheltered.

Indicators of disturbed aquatic systems in western North America may include: Largemouth Bass (*Micropterus salmoides*), Brown Bullhead (*Ameiurus nebulosus*), American Bullfrog (*Rana catesbeiana*) and Green Frog (*Rana clamitans*). Indicators of disturbed terrestrial systems across North America may include: European Starling (*Sturnus vulgaris*), House Sparrow (*Passer domesticus*), Domestic Cats (*Felis domesticus*), Norway Rat (*Rattus norvegicus*), Purple Loosetrife (*Lythrum salicaria* L.), Policeman's Helmet (*Impatiens glandulifera*), Himalayan Blackberry (*Rubus discolor*), Japanese Knotweed (*Polygonum cuspidatum*) and Scotch Broom (*Cytisus scoparius*).

Native Versus Non-native Biodiversity

Plants and animals evolve over time in response to their physical and biological environment. As such, native vegetation are able to support 50 times more native wildlife species than exotic plants (Pettinger 1996) due to the fact that native species have adapted together over geologic time. Native plants provide familiar sources of food and shelter for wildlife and they also require less external care in terms of fertilizer or pesticide use than exotic species. However, greater amounts of exotic ornamental vegetation are currently found in urban areas (Pickett et al. 2001). In addition, habitat fragmentation caused by urban development have also attracted introduced species that are able to adapt and exploit these disturbed environments unsuitable for most native wildlife.

What effects do exotic species have on biodiversity? This depends on the exotic species in question. Generally, exotic vegetation provide less habitat value for wildlife than do native. However, local wildlife are beginning to adapt. For example, most native pollinators have been observed to use introduced, weedy vegetation as nectar sources and host plants in addition to native stock. Blackberry thickets also provide great protection cover and food for many urban wildlife species, such as sparrows, finches, cottontail rabbits, raccoons and coyotes (Cacik and Schaefer 1998). On the other hand, blackberry thickets are invasive and often displace native plant species that other wildlife creatures depend upon for habitat value.

Problems arise as exotic species invade natural habitats. Blackberry thickets are invading riparian areas and purple loosestrife are altering wetland habitats.

Japanese knotweed is another exotic species that has changed the natural environment by clogging waterways and disrupting groundwater flows. Separated from their natural predators, exotic species are able to propagate, spread and exploit various habitats with little competition. Economic costs occur as invasive species take away the "free" ecological services that were provided by natural habitats before invaders disrupted their prior structure and function.

Minimum Habitat Requirements

Planning to conserve biodiversity and the complex ecological functions and relationships it supports requires a precautionary approach. Further research and closer familiarity with each municipal jurisdiction should look at adjacent land use and habitat quality within and surrounding each ecosystem to prioritize which areas are of primary concern and which areas have a better chance of rehabilitation or conservation. The use of indicator species can assist municipalities in assessing habitat quality and biodiversity.

Wetland Ecosystems

Several species assemblages utilize permanent and seasonal wetland ecosystems for all or part of their life cycle. Activities such as dredging, draining and filling are seriously depleting the quality and presence of wetlands in North America. A pattern of fewer and more isolated wetlands is occurring as urbanization and human density increases (Gibbs 2000). The dispersal abilities of wetland animals are thus affected. Average distances for many wetland animals are generally <0.3 km for frogs, salamanders and small mammals and <0.5 km for reptiles (Gibbs, 2000). In Gibbs' study (2000), preserving a wetland that is as small as 0.5 hectares in size and that is 500 m from other wetlands can retain metapopulations of most wetland organisms over the long term. Very isolated wetlands must be larger. Also, to conserve larger wetland species that have a larger home range size, more adequate wetland habitat must be protected.

Adjacent habitat types, such as rivers and streams and riparian buffer zones, are also necessary for protection as their quality and removal can impact and impair wetland functions. Removal of riparian habitat through development can also disrupt the natural cycle of nutrients flowing into wetland ecosystems. This can lead to nutrient starvation and drought as pavement prevents water from being absorbed by the Earth (Nowlan & Jeffries 1996). In addition, if smaller tributaries are dammed, culverted or drained, this can severely affect the survival of wetland ecosystems; therefore, even though the main part of a wetland is protected, the wetland and all its functions can still be lost if measures are not put in place to ensure the health of adjacent habitat types. Many terrestrial amphibians also utilize adjacent moist forested ecosystems during non-breeding seasons. Healthy water quality found in wetlands is essential if they are to be used as breeding sites. Adequate riparian vegetation offering shade, foraging opportunities and shelter are also required in healthy wetlands.

Vernal pools are favoured as breeding sites by several amphibian species, such as the Pacific Treefrog, since its temporary nature and shallowness inhibit use by predatory fish. Also, due to their shallowness, vernal pools are often warmer and this could also enhance larval growth (Walls et al. 1992). Amphibians using vernal pools pay a price as they face the threat of desiccation during egg and larval development (Hecnar & M'Closkey 1997a).

Open Water Ecosystems

Streams and ponds are permanent habitats for fish, many aquatic insects and some amphibians. A high level of water quality is crucial to sustain aquatic species diversity, which many terrestrial species also rely upon for food. To maintain these conditions, a riparian buffer offering distance from disturbances is necessary.

Several species, such as birds and amphibians, require habitat heterogeneity during their life cycle. Therefore, open water ecosystems adjacent to forested or herbaceous/grassland ecosystems are desirable, especially for residential and migrating shorebirds, waterfowl and seabirds to forage in.

Forest Ecosystems

To maintain or restore biodiversity in forests, their structural diversity and integrity must be upheld. For example, decaying wood provides habitat for numerous vertebrates, fungi, invertebrates, lichens, plants and micro-organisms. Coarse/downed woody debris and forest floor litter can create environments advantageous for fossorial (ground hiding) amphibians and small mammals. Snags provide complexity and added habitat opportunities for several indicator species, such as the Pileated Woodpecker (*Dryocopus pileatus*) and the Red-breasted Nuthatch (*Sitta canadensis*). Their home range is actually dependent on the density of snags in a forested area. If food sources are scarce, they will extend their home range.

Recommendations from other studies indicate that urban woodlots and forested areas should be a least 5 hectares, but optimally 20 to 30 hectares, to manage land vertebrate and bird species. Smaller and more isolated woodlots are also valuable provided they have a dense vegetative cover to maintain a moist microhabitat required for amphibians and insects. Habitat patches at least 0.55 hectares in size with sources of water are important for retaining amphibians and reptiles. Areas larger than 10 hectares with some clearings can also create conditions for forest edge and interior species. Many forest interior bird species, such as Brown Creeper (*Certhia Americana*), have never been observed in woodlands smaller than 5 hectares (Adams & Dove 1994).

Riparian Ecosystems

Riparian ecosystems provide many functions. Species that rely on both aquatic and terrestrial habitats during their life cycle, such as the northern river otter and some amphibians, greatly benefit from the shelter, cool shade and foraging opportunities that streamside vegetation provides. Riparian vegetation also helps sustain high water quality conditions, which contributes to the health of aquatic species diversity. A high level of structural plant diversity consisting of snags, trees, thick shrub layers and a somewhat moist understorey should be maintained. For example, within the Forest Practices Code of British Columbia, 30m on either side of a non-fish bearing, permanent stream (>3m channel width) has been assigned as a minimum setback distance protected from development (Ministry of Forests 1995a).

More Marginal Disturbed Systems

Although urban areas have resulted in habitat islands, degradation and loss, we cannot overlook the natural habitats, though small and fragmented, found in cities, nor can we discount them from any responsibility in biodiversity conservation. The actual potential of green space expansion or acquisition may be restricted, but the quality and restoration of barren landscapes, monoculture utility right of ways, manicured urban parks, and backyard, balcony or rooftop space can be improved to provide greater value to wildlife and, thus, increase the level of urban biodiversity. The discontinuation of the use of chemicals that affect the survival and habitat quality for wildlife species (including ourselves) would ensure healthier water systems. Increasing the concentration of native vegetation can also decrease the amount of fertilizer and pesticide needed to sustain ornamental exotic plant species.

Planting riparian areas helps maintain the biodiversity of streams.

Backyard ponds provide valuable breeding habitat for Pacific tree frogs (*Hyla regilla*), which can easily take advantage of standing water pools with little threat of predation. American Robins have also been known to nest within potted plants placed outside balconies in apartment buildings. The simplest of initiatives undertaken by local citizens can create significant results in terms of biodiversity. Providing food, shelter, space and a source of clean water, and replacing grass with native shrubbery can provide valuable habitats for insects, birds and amphibians.

In rural areas, wildlife can benefit from structures such as hedgerows, shelterbelts or old fields. Songbirds, raptors and small mammals, such as voles, take advantage of these habitat types. Nearby clean water is another important habitat requirement for rural areas.

Herb and Grass Ecosystems

Old-fields that have a moderately dense grass cover between 25–50 cm high with a litter mat between 10–15 cm deep provide an ideal breeding space for Townsend's voles (*Microtus townsendii*) (Butler 1992). Old-fields also attract wintering raptors such as Short-eared Owls (*Asio flammeus*), Rough Legged Hawks (*Buteo lagopus*) and Northern Harriers (*Circus cyaneus*) because of the good visibility and amount of prey they offer. Maintenance of this habitat through periodic mowing should keep fields at an optimum stage of field succession. Undesirable plant species, like blackberry patches should be removed, since these plants diminish the value of old fields as feeding areas for vole-eating raptors.

Pastures and croplands surrounded by hedgerows and shelterbelts also provide a wide range of habitat needs for songbirds and small mammals. During winter nights, huge flocks of migrating waterfowl make use of pastures for grazing and foraging opportunities. The American Wigeon (*Anas americana*) and Mallard both obtain much of their food energy from grass and pasture in farmlands. These areas are also a forage ground for juvenile Great Blue Heron (*Ardea herodias*) (Butler 1992).

Tall grasses are an optimum habitat for the Townsend's vole and its predators.

Athletic fields and golf courses are short grass fields. Airports and cemeteries also represent this habitat. Many species of shorebirds, such as Canada Goose (*Branta canadensis*) and American Wigeon graze in this habitat type (Acres International Ltd. 1993). They are also popular feeding grounds for passerines, such as the Barn Swallow, American Robin (*Turdus migratorius*), House Finch (*Carpodacus mexicanus)* and European Starling, especially if standing water sources, like ditches, dykes or ponds are located nearby. Golf courses can also provide some habitat value if small forested patches are left to grow and understorey shrubbery is allowed to develop. This will allow amphibians, insects and songbirds to find their niche. Short grass fields are not the ideal habitat, but they are better than paved lots in terms of habitat value.

Built Environments

Built environments (i.e., bird boxes) mimic natural conditions that are critical for the survival of some species that have been displaced from their natural nesting sites due to habitat loss or invasive species. While there is no minimum size or number for wildlife enhancements, they do reduce impacts of habitat loss, support wildlife and provide recreational viewing opportunities.

Buildings can be compared to cliff or bluff environments. Both are harsh, but capable of providing specialized habitat for wildlife. Crevices, openings and designed holes or pockets within buildings can function as shelter for bats and birds, including the Barn Owl (*Tyto alba*). The amount of habitat value is limited, but can be enhanced through container plantings, balcony gardens and creation of green space surrounding paved environments to serve as a sink for songbirds, such as Spotted Towhees (*Pipilo maculates*) or American Robins which tend to avoid vacant lots normally occupied by crows and starlings.

Exposed and Disturbed Sites

Exposed and disturbed sites are found throughout the Greater Vancouver area. In quarries and gravel pits, the vegetation present usually consist of moss and lichen growing on rock surfaces and herbaceous plants and grasses that have been introduced to the area through seed dispersion by wind or rain. Patches of clover and alder stands could develop, attracting bees and songbirds. However, the wildlife habitat provided by quarries and gravel pits are minimal. Through rehabilitation, these areas could become productive wildlife habitats.

A water-filled ditch will greatly increase the number of species using grass fields.

Habitat Fragmentation and Loss

As cities grow, the natural habitat is forced to accommodate. This has led to the creation of more isolated fragments of suitable living space for wildlife (Adams & Dove, 1994). Edge effects are a result of habitat fragmentation. Although they can occur naturally and be caused by landscape features related to topographic differences, soil type, presence of open water, or geomorphic factors (Bannerman, 1998), induced edges are created by humans when forests are logged, or when the natural habitat is disturbed to make room for urban development. Patch sizes are reduced and patch edges have greater exposure to sun and wind, which can cause microclimatic changes. Drier edges can alter growing conditions for interior plant and animal species (Lehmkuhl & Ruggiero 1991). These microclimatic edge effects can extend 160–200 m into a forested patch (Lehmkuhl & Ruggiero, 1991; Bannerman 1998).

Species richness generally increases with habitat fragment size up to a certain point. Larger fragments showed a decline in species richness due to the loss of "edge" species and establishment of shade tolerant vegetation (Adams & Dove 1994). Wildlife sensitivities toward habitat fragmentation are dependent on habitat specificity and dispersal abilities (With & Crist 1995). For species that rely on more than one habitat type, migrate, or require a large territory, the spaces between these fragments are difficult obstacles to survive and overcome. Adjacent habitat and land use also affect species composition (Pickett et al. 2001). In addition, the "edge contrast", or how sharply two adjacent habitats differ, helps determine how strong edge effects are on wildlife species (Bannerman 1998).

Amphibians and Reptiles

Amphibians require moist environments, preferably close to pristine water bodies. Many amphibians migrate between aquatic and terrestrial habitats during their life cycle. The health and quality of these habitat types are generally associated with mature forest interiors, riparian areas and wetlands. Habitat fragmentation reduces the size and quality of these areas by creating drier edge habitats and by exposing waterways to heat and potential urban runoff and sedimentation.

Clearcutting, for example, can decrease the amount of large downed wood available for amphibians to nest in and hide beneath. It also can reduce canopy and vegetation cover and cause soil compaction, which can restrict amphibian forage activities and movement patterns (Dupuis & Waterhouse, 2001). Plethodontids, such as the red-backed salamander, are completely terrestrial and they rely heavily on animal burrows, root channels and rock fissures to move between the moister depths of the soil, talus and forest surface (Welsh and Droege 2001). As riparian vegetation is removed during clearcutting, this can also reduce stream and pond habitat quality through sedimentation, increased water temperatures and accelerated water loss.

Dredging and damming of shallow, ephemeral wetlands have replaced this valuable habitat type with smaller, warmer and more vegetated permanent ponds (Kiesecker et al. 2001). These activities not only fragmented the original habitat, but also created a new habitat favourable to introduced species such as the bullfrog and green frog; these species can occupy niches unfavourable to other native amphibians, such as the red-legged frog (*Rana aurora*) and displace them (Kiesecker et al. 2001). More information regarding the bullfrog and green frog can be found in Section 5.1.3 and 5.1.4.

Roads also appear to be an important landscape feature that hinders amphibian movements. Forest-road edges are much less permeable than forest edges associated with open lands or residential areas (Gibbs 1998a). This could be due to a behavioural avoidance of road edges or elevated mortality associated with road crossing (Gibbs, 1998a).

In a study conducted by Gibbs (1998), the dispersal ability of amphibians was inversely related to fragmentation resistance. Mobile species were more vulnerable to habitat fragmentation than more sedentary species. Mobile species were more likely to end their migration in unsuitable habitats or become stranded in open or vulnerable sites between forested areas. This subsequently resulted in failure for them to reach their breeding habitat to propagate. Those species that did not migrate or disperse as far were less affected by habitat fragmentation (Carr & Fahrig 2001). However, if amphibians with limited dispersal capabilities are trapped in a patch area that is unable to fulfill their habitat requirements, they are also vulnerable to population declines.

Birds

Many resident and migratory bird populations in the Greater Vancouver area are habitat specialists. They are, therefore, also susceptible to the negative effects associated with habitat fragmentation, such as habitat loss and degradation. Bird species at higher risk of habitat fragmentation are generally larger and have high mobility. Larger birds generally have larger home ranges than smaller species and also expend more energy to move (Lehmkuhl & Ruggier, 1991). Smaller patch sizes would thus require them to expend more energy to reach other suitable patch habitats to find food and shelter. The smaller and more isolated habitat fragments are, the more energy larger birds will need to expend. This can result in failure to find suitable nesting areas or adequate food sources. The Great Blue Heron, for instance, relies on mature forests, healthy grasslands and aquatic (wetland and open water) ecosystems in relatively close proximity to each other to fulfill their habitat requirements (Butler, 1997).

Edge habitats have also been referred to as an "ecological trap" (Harris 1988), due to the abundance of edge-related predation that can occur. Many ground feeding birds, such as the Song Sparrow (*Melospiza melodia*), Killdeer (*Charadrius vociferous*) or American Robin, may be attracted to forest edges due to the variety of vegetation and abundance of food. However, edge-related nest predation caused by small mammals, snakes, ravens, crows and cowbirds have all been cited as causes for population declines in birds (Bannerman, 1998; Harris, 1988). Many birds, such as the Pileated Woodpecker, have traditionally avoided heavy levels of nest predation by nesting in the interior of large forest tracts (Harris 1988). Pileated woodpeckers require contiguous forest to sustain their breeding population. Edge-related predation may extend as far as 600m into the interior habitat (Bannerman 1988).

Many generalist bird species make use of edge habitats and urban areas. Two examples are the European Starling and the House Sparrow.

Mammals

Mammals that require a large home range are vulnerable to habitat fragmentation. Black bears require large tracts of relatively undisturbed forest and often will not cross large areas of non-forested land to reach other suitable habitat patches (O'Neil et al. 2001). Some black bears that inhabit the forests of the north shore of the Greater Vancouver Region have been forced into urban areas in search of food due to increased habitat fragmentation. Larger mammals, including coyotes, pose safety risks to the human population and unfortunately force is sometimes used to control or eliminate the problem that habitat fragmentation has created.

Fragmentation affects the environment by reducing the availability and heterogeneity of original habitats. Smaller mammals that require moist forested habitats with structural diversity are negatively affected by exposure to edge effects. The red-backed vole (*Clethrionomys sp.)* and Douglas squirrel (*Tamiasciurus douglasi*) are two small mammals that require the structural diversity associated with forested habitats. Eliminating habitat for some larger predators has been an advantage for smaller predators, which can increase in numbers and decimate prey species that formerly persisted with low populations of small predators.

Northern river otters are also sensitive to habitat fragmentation. These mammals live in both riparian and aquatic habitat, and usually do not venture farther than 200 m from water. Their habitat use also tends to be linear in nature. Through the habitat fragmentation associated with urban development, river otters are forced to alter their home range size and habitat use patterns to adjust to this fragmentation.

Insects

Habitat fragmentation effects on bees and other insects are mixed. In some cases, habitat fragmentation has led to an increase in insect and bee diversity. These insects are able to capitalize on and exploit the weedy plant species that seem to proliferate in edge habitats and disturbed areas. However, in other cases habitat fragmentation caused by logging or urban development has altered drainage patterns, caused soil compaction and altered habitat requirements to the point that insects and other wildlife animals are displaced. For example, the bumble-bee is a ground nester. A suitable nesting habitat requires soil of specific texture, depth, slope, vegetation cover and moisture. The reduction of these conditions caused by habitat fragmentation and degradation can lead to a decline in population and pollination activity (Allen-Wardell et al. 1998). Most bees of medium size have a home range of 1–2 km. However, the distance between forage patches is as important as permeability surrounding these habitat fragments (Cane 2001); therefore, adjacent land use is also an important consideration when determining its value for flying species.

Plants

Edge habitat conditions are usually drier, facilitating colonization by vigorous seral plant communities and associated wildlife species (Lehmkuhl & Ruggiero 1991). Some hardy weed species that can tolerate extreme temperature changes in open areas are very adaptable to edge habitat. Being prolific seed dispersers, winds can take these seeds and deposit them within the habitat patch, eventually altering the entire composition of the interior's flora (Bannerman 1988). Ground layer vegetation that is intolerant to trampling and soil compaction is also sensitive to habitat fragmentation and urban development (Adams & Dove 1994).

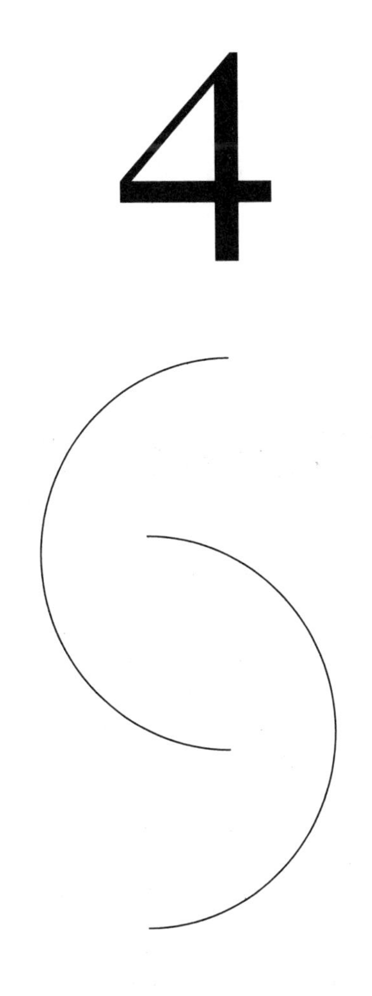

4 Urban Ecosystem Services

Urbanization has changed the landscape. In place of forests, grasslands, and wetlands are now buildings, asphalt, and concrete. Streams are covered and culverted, parks are made up of turf grass, and mountain hillsides are large subdivisions. This has drastically changed the ability of the natural processes to cycle nutrients and water and keep the Earth clean. As cities grow we have had to replace the services provided by nature with expensive technologies to purify water, purify air, treat sewage and burn garbage.

These ecosystem services are still often taken for granted, partly because we do not understand them and partly because they are very hard to quantify and assign a value. However, they have caught the attention of many people concerned about urban planning and creating livable regions. These services are integral to maintaining the quality of life for people in the city and are now considered to be part of its infrastructure, just as are roads, energy supply and waste management. This green infrastructure is a life support system, including a network of agricultural lands, parks, greenways, wilderness areas and easements (www.greeninfrastructure.net/). The green infrastructure is strategically planned and managed to conserve native species and maintain natural ecological processes in order to maintain the environment of air, water and land.

This chapter will examine some of the general evaluations that have been used to estimate the value of urban ecosystem services. Chapter 5 that follows contains a more detailed treatment of how nature's services can be calculated for small scale plantings typical in cities.

Nature's Services

Nature's services — also known as ecosystem services — are the conditions and processes provided by nature that enable human survival (Daily 1997). In other words, they are "the benefits human populations derive directly or indirectly from ecosystem functions" (Constanza et al. 1997). These services are the goods, such as food and raw material production, and services, such as nutrient cycling and disturbance regulation that are provided by natural systems.

The following are examples of nature's services (Daily 1997):

- Purification of air and water
- Natural flood control
- Disposal of waste and dead organic matter in soils
- Detoxification and decomposition of wastes
- Pollination of agricultural crops and natural vegetation
- Dispersal of seeds
- Regulation of nutrient cycling through ecosystems
- Control and prevention of soil erosion
- Control of pests by natural predators and parasites
- Detoxification of wastes
- Dilution of pollution
- Maintenance of biodiversity
- Protection of coastal shores by erosion of waves
- Protection from the sun's UV rays
- Climate regulation

Nature's services are often taken for granted, forgotten about completely, or entirely unknown. One of the reasons these services are forgotten is that we don't actually pay for them. The monetary value of ecosystem goods is not accurately represented when considering all the organisms and processes that go into producing them. For example, timber has a value on the open market and is sold throughout the world, but the cost of the soil organisms that contribute to soil fertility and the nutrient cycles that help the tree grow are not included in its price. Some attempts of valuation have been done and estimate these services to be worth an average of $33 trillion US per year globally. As a comparison, the global gross national product is only valued at US$18 trillion per year (Constanza et al. 1997).

Nature's services are largely irreplaceable with no synthetic or human-made substitutes. Natural gas and climate are examples of the many services for which there are no human-made substitutes on the scale found in nature. If ecosystem function were to become compromised, many services provided by these systems would also decline (Daily 1997).

Ecosystem services are extremely complex and interconnected. They operate on a variety of scales and some services require others in order to function. The problem is that we do not generally view the world as an interconnected ecosystem that requires all parts (both large, small, and unknown) to function. Most people do not associate soil and the ecosystem services it provides with the food they put on their table each day (Daily 1997).

Soil, Plants, and Pollination

Soil is a very diverse and dynamic ecosystem type. It is a central part of all food webs and is an indispensable component of life. Soil provides six major

Figure 4.1 The Interconnectedness of Soil, Plants and Pollination

If all the plants were cut down, there would be nowhere for all the insects and wildlife that depend on vegetation to live. There would also be nowhere to get the medicinal ingredients that come from plants. Many insect pollinators would have nowhere to live, as well as no plants to pollinate. Soil would be lost because the micro-organisms would be lost and the decomposition of dead plant material would not occur. There would be no recycling of nutrients; therefore, no new plants. There will be no place for wildlife (birds and mammals) to find shelter, food, or nesting sites.

ecosystem services: buffering and moderation of the hydrological cycle, physical support of plants, retention and delivery of nutrients to plants, disposal of wastes and dead organic matter, renewal of soil fertility, and regulation of major element cycles (Daily 1997).

While soil provides physical support for plants, plants help prevent soil erosion by stabilizing the landscape. Vegetation also helps to regulate water flows and helps prevent flood and drought regimes. Plants moderate climate by regulating rainfall regimes and the albedo effect (the absorption of heat from the sun on the Earth's surface and its subsequent radiation back into the atmosphere) (Daily 1997). Global warming is regulated since plants, especially trees, act as a carbon sink. Vegetation also provides food, fuel, and timber. Plants are also biologically diverse and support a large amount of animal diversity by providing wildlife habitats.

Plants play an important role in the medical and pharmaceutical industries. One in four of our medicines and pharmaceuticals is derived from plants (Daily 1997). For example, a compound in the bark of yew trees (*Taxus spp.)* is effective at fighting various forms of cancer. There are also two drugs derived from Madagascar's rosy periwinkle (*Vinca sp.)* plant that are used against childhood leukemia, Hodgkin's diseases and other cancers. Tens of thousands of lives are saved every year in North America by disease-fighting drugs derived from plants.

Both cultivated food plants and wild flowering plants rely on pollination for survival. Bees, other insects, bats, birds, and wind are responsible for pollinating plants. Insect populations throughout the world are declining due to diseases and parasites, such as the mite responsible for declining honeybee populations in North America in the late 1990s. The widespread use of herbicides, insecticides and other chemicals on cultivated plants also kills many beneficial insect pollinators. Declining pollinator populations are of concern since over 95% of all cultivated food plants rely on insect pollination (Daily 1997).

Water and Wetlands

Freshwater ecosystems provide many services. These include water for drinking, irrigation, manufacturing, recreation, transportation, flood control, bird and wildlife habitat, and dilution of pollution.

Currently only 0.77% of the world's available water supply is freshwater that is not locked up in glaciers (Daily 1997). Through urban development, the amount of freshwater available as groundwater is decreasing due to storm water management. When it rains, water runs off into the storm water/sewer system. These sewers flow into a creek's system that then flows into the ocean. Therefore, this water is not available to replenish groundwater systems and little water is available for plants. Storm water mismanagement also causes flooding in streams due to the influx of water during periods of heavy rainfall and little to no water input during periods of little rain. Flooding wipes out insects, fish eggs, and vegetation in creek beds. If no salmon eggs survived a flood event, there would be no salmon as food for other marine life, such as whales, seals, and others.

In the Amazon, a significant percentage of the mean annual rainfall is recycled by the forest through evapotranspiration (water evaporating from plant leaves). If there were no plants, or if plant cover was reduced through deforestation, total precipitation would be greatly reduced and vegetation would be unable to re-establish itself (Daily 1997). Not only would there be a decline in the overall biodiversity, the region would become much drier, changing the natural environment and climate.

As with many natural systems, wetlands are extremely diverse and ecologically important. They provide a variety of ecosystem services, including evolution of unique species, production of harvested wildlife, production of wildlife for nonexploitative recreation, production of wood and other fibres, water quality improvement, flood mitigation and abatement, water conservation, carbon accumulation, methane production, denitrification, and sulphur reduction (Daily 1997).

Interconnections

The previous examples provide a base for understanding what nature's services are and how they depend on other services or ecosystems for their continued existence.

Nature's services are like the making of a fine cake (Daily 1997). A baker will share the ingredients and knowledge of how to transform them, describing the oven, pan, and other appliances required. If pressed further, the baker might point out the need for capital, infrastructure, and human services to process, store, and transport the ingredients. If prompted, the baker may mention the cropland, water, chemical, and energy inputs, but will probably forget to tell you about the natural renewal of the soil fertility, pollination of crops, natural pest control, the role of biodiversity in maintaining crop productivity, clean up, and recycling services. All of these things are part of the recipe as well.

Some services are more relevant to urban ecosystems, such as air filtering, micro-climate regulation, noise reduction, rainwater drainage, sewage treatment, erosion control and recreational and cultural value. However, we must keep in mind that while we live in the city, we use services located outside the city. Some research estimates that urban centres require areas 500 to 1000 times larger than the city itself to be sustained (Bolund and Hunahammer 1999). These areas outside the cities are used to grow food and other products essential to everyday life in the city.

Part of understanding the services we use is to value them and try to reduce our impact on natural ecosystems. Throughout the remainder of the book, we will explore urban ecosystems and the services they provide. We will also provide suggestions on how you can help improve urban ecosystem services.

Removing Carbon Dioxide from the Atmosphere — Carbon Sequestration

Climate change, a change in the average weather conditions of a region, has garnered a lot of media attention in the past few years. Climate change has happened naturally since the beginning of time and has led to ice ages and very warm periods. This occurs because the Earth naturally acts as a greenhouse. Gases, such as water vapour, carbon dioxide and methane, trap the heat from the sun in the atmosphere. The current concerns are the amount of carbon dioxide, the predominant greenhouse gas, in the atmosphere and the rate at which it is accelerating. It is believed that carbon dioxide will double in the next 50 years and that this influx of carbon dioxide is the result of human activities, mostly automobiles and deforestation (Nowak 1993). The increase in carbon dioxide is causing the surface of the Earth to heat up more quickly than in the past. Even a small change in temperature on the Earth's surface could lead to dramatic differences in the biological world. Ice caps could melt, flooding coastal areas. Biogeoclimatic zones could shift, having a marked effect on our farmland's viability and our ability to produce food. Animals and plants that may historically have once moved with gradually shifting zones will be cut off due to urban development and the rapid pace of this warming period. This will undoubtedly lead to mass extinctions.

Total atmospheric carbon is estimated to be increasing by 2.6 billion tonnes every year (Nowak 1993). In the Greater Vancouver Regional District (GVRD) in 1998, 15,334,000 metric tonnes of greenhouse gases were added to the atmosphere, or 7.6 tonnes per person living in the region (www.gvrd.bc.ca June 15, 2001). The average American contributes about 10 tonnes of carbon dioxide to the environment each year (www.americanforests.org June 15, 2001). To calculate the amount of carbon dioxide you personally might be contributing to the environment each year, use the carbon calculator on the American forests website (www.americanforests.org).

Since trees sequester carbon in their growth process, taking the carbon from carbon dioxide and converting it to carbon in sugars and cellulose, they can act as a temporary sink for this atmospheric carbon (Nowak 1993); note that a tree will release the carbon back to the atmosphere when it dies and decomposes so the benefit is limited to the lifespan of the tree. Trees not only store carbon but also have the added benefit of being able to help avoid carbon production through energy conservation.

Much research has been done on forests and carbon sequestration rates, but the urban environment is quite different. There have only been a few studies to quantify the amount of carbon stored in urban forests. Trees in urban environments are subject to different kinds of stresses, such as polluted soil and air (Johnson and Gerhold 2001). Urban trees also tend to grow in more open space, which leads to trees that are generally shorter with larger crowns and more branches than forest trees (Nowak 1994). This can lead to different growth characteristics among trees of the same species. Urban forests typically store only half the amount of carbon dioxide stored by natural forests (McPherson 1998). The amount of carbon that can be stored is influenced by the extent of existing tree canopy cover, tree density, and pattern of tree diameters in the city (McPherson 1994). For example, Oakland, California, has 21% tree cover and stores 11 tonnes of carbon/ha of land, and Shorewood, Wisconsin, has 39% tree cover and stores 23 t/ha (Nowak 1994). The amount of carbon a tree can sequester also depends on its size. The amount of carbon stored in Sacramento, California's urban forests is 172 t/ha, which is due to Sacramento's high percentage of large diameter trees compared to other cities (McPherson 1998). These estimates do not account for carbon stored elsewhere in the environment, such as soils, shrubs, and grasses (Nowak 1991). The average carbon storage by trees in Chicago, Illinois, which has 11% tree cover, is 14 t/ha; however, this average varies greatly depending on the diameter of the tree (Nowak 1994), with smaller trees storing less carbon than larger trees.

The amount of carbon sequestered each year by a tree in Chicago that is less than 8 cm diameter at breast height (dbh) is approximately 1 kg/year whereas a tree greater than 76 cm dbh sequesters about 93 kg/year (Nowak 1994). Therefore, not only do large trees store more carbon overall, they also are able to sequester more carbon per year than smaller trees. Smaller trees store most of their carbon in their trunk but, as dbh increases, the amount of carbon stored in the trunk compared to the branches decreases by almost half (Johnson and Gerhold 2001). In order to keep carbon out of the atmosphere, it is important to reduce the amount of pruning required. It is also important to keep larger trees healthy and continually replace trees that have died so there will always be older, larger trees able to take up more carbon dioxide. The financial benefits of this ecosystem service vary greatly between studies. In Sacramento, carbon sequestration is valued at $0.55 per tree/year in avoided carbon dioxide costs (McPherson, 1998). In Modesto, California, it was estimated at $3.94 per tree/year (McPherson et al. 1999). In Chicago, carbon uptake was estimated at $1.03 per tree per year for the average 30-year-old green ash tree (McPherson 1994b).

Air Filtration

Air pollution is a major social, economic and environmental concern in most cities. It is estimated that 60,000 people die each year in the US from the adverse health effects of air pollution (Nowak 1994b). In the next 25 years, it is estimated that emissions reductions in British Columbia's Lower Fraser Valley will prevent 2,000 premature deaths and 27,000 emergency room visits, and have an economic net benefit of C$2.5 billion (www.gvrd.bc.ca, June 15, 2001). Air quality is also very important to the Greater Vancouver Regional District because one of the largest industries is tourism. In 1998 7.9 million visitors to the GVRD spent $2.7 billion in the area (Tourism Vancouver 1998 Annual Report). One reason tourists are so attracted to this area is the natural environment and the outdoor recreation opportunities. An important part of that natural environment is good air quality.

Air pollution not only affects health but also damages vegetation and buildings, reduces visibility, and contributes to acid deposition (Nowak 1994b). Trees reduce atmospheric pollutants by filtering particles, absorbing gaseous pollutants, reducing city temperatures, and reducing energy building consumption (Nowak 1991).

Major air pollutants include carbon monoxide, nitrogen oxides, ozone, sulphur oxides, and particulate matter (PM_{10} refers to particulate matter less than 10 microns in size).

- **Carbon monoxide** comes predominantly from automobiles and is harmful because it binds with the hemoglobin in blood, reducing the amount of oxygen that it can distribute through out the body.

Beautiful scenery is what attracts people to some major cities. The air, water and soil pollutants we create are removed or detoxified by the living process of natural systems that rely on biodiversity to function.

- **Nitrogen oxides** (the main one being nitrogen dioxide) come from automobiles and stationary combustion engines and can have negative effects on plants by forming nitric acid.
- **Ozone** is formed through chemical reactions in sunlight (UV rays) involving nitrogen oxides and volatile organic compounds. Ozone is harmful to living tissue, damaging lungs and plant leaves.
- **Sulphur oxides** (mainly sulphur dioxide) come from stationary combustion sources and the smelting of ores and can have some negative effects on plants but at quite high levels by forming sulphuric acid (Nowak 1994b).
- **Particulate matter** results from soils, industrial processes, combustion products, and chemical reactions involving gaseous pollutants. These particles pose significant health problems as they escape the defence mechanisms of the respiratory tract (the hairs and mucus that trap and hold foreign particles) and

Table 4.1 Air Pollution and Health

Pollutant	Health Effect
Nitrogen dioxide	• Aggravation of asthma and allergy symptoms • Acute respiratory illness in children • Increased airway resistance and reduced lung function
Sulphur dioxide	• Bronchoconstriction • Also associated with acute morbidity and elevated mortality in epidemiology studies, although collinearity with particulates is suspected
Ozone	• Aggravation of asthma and other chronic respiratory diseases • Lung function reductions and respiratory irritation • Acute respiratory illness
Total particulate matter	• Elevated risk of mortality • Lung cancer • Higher prevalence of chronic respiratory disease • Acute illness, including work loss and emergency room visits • Lung function reductions and respiratory symptoms

(Source: Chestnut and Rowe 1989)

enter into the lungs. Particles smaller than five microns may settle in the alveoli of the lungs for weeks to years (Nowak 1994b).

Table 4.1 offers a small sample of some of the human health risks associated with these pollutants.

More than 2,000,000 people live in the 400,000 hectares that make up the Greater Vancouver Regional District (www.gvrd.bc.ca June 15, 2001, Wackernagel and Rees 1996). In 1998, Greater Vancouver added into the atmosphere: 191,000 tonnes of carbon monoxide (CO), 48,000 tonnes of volatile organic compounds (VOCs), 42,000 tonnes of nitrogen oxides (NOx), 4,000 tonnes of sulphur oxides (SOx), and 16,000 tonnes of particulate matter. Also added to the atmosphere were 15,334,000 tonnes of greenhouse gases (such as carbon dioxide, ozone, and methane) (www.gvrd.bc.ca June 15, 2001).

In the GVRD, each person is responsible for 95 kg of CO, 24 kg of VOCs, 21 kg of NOx, 2 kg of SOx 8 kg of particulate matter, and 7,606 kg of greenhouse gases annually. Many people believe industry is the cause of most of our air problems but that is not the case; 75% of these emissions are from motor vehicles — 58% from passenger vehicles and 17% from transport vehicles (www.gvrd.bc.ca June 15, 2001). The good thing about this situation is that it is changeable. We can make different lifestyle choices and plant more vegetation to help mitigate the air pollution.

Plants can take up some of these pollutants without harm and provide a valuable ecosystem service by cleaning the air. The rate at which trees remove pollutants depends on the amount of foliage, the number and condition of stomata, and meteorological conditions (Dwyer et al. 1992). In Chicago (11% tree cover) in 1991, trees removed 15 tonnes of carbon monoxide, 84 tonnes of sulphur dioxide, 89 tonnes of nitrogen dioxide, 191 tonnes of ozone, and 212 tonnes of PM_{10} (Nowak 1994b). This service was valued at $1 million. Across the broader area of Chicago, suburban Cook County and Dupage County trees removed 5,600 tonnes of pollution; this service was worth $9.2 million (Nowak 1994b). Removal rates were 9.7 kg/ha/yr for Chicago and 16.7 kg/ha/yr for the entire study area (Nowak 1994b).

In 1990, Sacramento trees removed 1,457 tonnes of air pollution at an estimated value of $28.7 million. The average uptake rate was 10.9 kg/ha/yr, which represents 1–2% of annual anthropogenic emissions (Scott et al. 1998). This study found that the approximate air pollutant removal benefit to be $5 per tree per year (Scott et al. 1998). In Modesto, California, the costs and benefits of trees were analyzed and it was found that the benefit of pollution uptake plus avoided emissions from power plants was $16 per tree (McPherson et al. 1999). The study in Chicago estimated air filtration and avoided emissions to be worth between $4.63 and $3.39, depending on where the tree was planted (McPherson 1994b).

Leaves of a vine maple. Many of nature's services can be estimated through a consideration of the leaf surface area represented by plants.

The differences in estimates of benefits of pollution uptake are dependent on which emissions control costs were used to determine the benefits. In Modesto, these numbers were estimated by the market value of pollutant emission credits for the area (McPherson et al. 1999). In Sacramento and Chicago, the values were estimated using emission control cost factors for the area (Scott et al. 1998 and McPherson 1994b).

The filtering capacity of vegetation increases with leaf area, so trees will filter more than shrubs, and shrubs more than grass (Bolund and Hunhammar 1999). Trees with dense crowns and hairy leaf surfaces give the most shade and dust interception (McPherson 1989). Large trees have the greatest impact due to their larger leaf surface area (Nowak 1994b). A study by Beckett et al. (2000) on the effectiveness of certain tree species to improve air quality showed that coniferous trees were very good at intercepting large amounts of air particles, partially because they have needles all year round, and that decid-

Table 4.2 Top rated tree species for improving air quality.

Ozone		Carbon Monoxide		Overall	
Scientific Name	*Common Name*	*Scientific Name*	*Common Name*	*Scientific Name*	*Common Name*
Ulmus procera	English elm	Tilia americana*	American linden	Ulmus procera*	English elm
Tilia europea*[I]	Linden	Fagus grandifolia	American beech	*Tilia europea*	Linden
Fagus grandifolia	American Beech	*Tilia tomentosa**	Silver Linden	*Liriodendron tulipifera**	Tulip tree
Betula alleghaniensis[I]	Yellow birch	*Ulmus rubra*	Slippery elm	*Metasequioa glyptostroboides**	Dawn redwood
Liriodendron tulipifera*[S]	Tulip tree	*Fagus sylvatica*	Common beech	*Fagus grandifolia*	American beech
Tilia americana*	America Linden	*Betula alleghaniensis*	Yellow birch	*Tilia platyphyllos**	Linden-Lime
Fagus sylvatica	Common beech	*Tilia euchlora**	Crimean Linden	*Betula alleghaniensis*	Yellow birch
Tilia platyphyllos*[S]	Linden-Lime	*Ulmus procera**	English elm	*Fagus sylvatica*	Common beech
Metasequioa gylptostroboides*	Dawn redwood	*Ginkgo biloba**	Maidenhair tree	*Tilia americana**	American linden
Betula papyifera	Paper birch	*Liriodendron tuplipifera**	Tulip tree	*Ulmus Americana*	American elm
				Ulmus thomas	Cork elm

* Species recommended for street tree use in urban conditions. Hardiness zone and other tree factors need to be considered in urban tree selections

I Intermediate tolerance to pollutant

S Sensitive to pollutant

T Tolerant to sulphur dioxide; unknown tolerance to nitrogen dioxide

I/U Intermediate tolerance to sulphur dioxide; unknown tolerance to nitrogen dioxide

S/U Sensitive to sulphur dioxide; unknown tolerance to nitrogen dioxide

T/S Tolerant to sulphur dioxide; sensitive to nitrogen dioxide

Table 4.2 Top rated tree species for improving air quality cont'd

Particulate Matter		Sulphur/Nitrogen Dioxide		Overall	
Ulmus procera*	English elm	*Ulmus procera**[I/U]	English elm	*Chamaecyparis lawsoniana*	Lawson cypress
Platanus occidentalis*	American plane tree	*Tilia europea**[T/S]	Linden	*Tsuga heterophylla*	Western hemlock
Chamaecypris lawsoniana	Lawson cypress	*Populus deltoids*[T]	Cottonwood	*Tilia cordata**	Little leaf linden
Cupressocyparis x leylandii	Leyland cypress	*Platanus occidentalis**[T]	American plane tree	*Tsuga mertensiana*	Mountain hemlock
Juglans nigra	Black walnut	*Platanus x acerifolia**[T]	London plane tree	*Tilia tomentosa**	Silver linden
Eucalyptus globules	Eucalyptus	*Metasequioa gylptostroboides**[T]	Dawn redwood	Betula papyrifera	Paper birch
Tilia europea	Linden	*Liriodendron tulipifera**[T]	Tulip tree	*Celtis laevigata*	Sugar hackberry
Abies alba	Silver fir	*Juglans nigra*[S/U]	Black walnut	*Fraxinus excelsior**	European ash
Larix decidua	European larch	*Betula alleghaniensis*[S]	Yellow birch	*Ulmus crassifolia*	Cedar elm
Picea rubens	Red spruce	*Fagus grandifolia*	American Beech	*Betula nigra**	River birch
				Larix decidua	European larch

*Species recommended for street tree use in urban conditions. Hardiness zone and other tree factors need to be considered in urban tree selections

I Intermediate tolerance to pollutant

S Sensitive to pollutant

T Tolerant to sulphur dioxide; unknown tolerance to nitrogen dioxide

I/U Intermediate tolerance to sulphur dioxide; unknown tolerance to nitrogen diox-ide

S/U Sensitive to sulphur dioxide; unknown tolerance to nitrogen dioxide

T/S Tolerant to sulphur dioxide; sensitive to nitrogen dioxide

(Adapted from Nowak 2000.)

uous species with rough or hairy leaf surfaces also intercepted more particles than smooth-leaved varieties. However, while conifers have more leaf area year round they are also more susceptible to pollution than deciduous trees (Bolund and Hunhammar 1999). Pollution-resistance would enhance trees' survival in the long term and allow the maximum amount of pollution to be filtered. Conifers should be planted to remove particles throughout the winter, but thought should be given to site and species selection (Nowak 1994b).

Trees can contribute to air pollution as they emit VOCs, which act as precursors for ozone formation. However, since ozone formation is temperature-dependent and trees generally lower the air temperature, it is believed that increased tree cover lowers the overall VOCs and ozone levels in the urban environment (Nowak et al. 2000).

Energy Reduction

There is a need to cool our cities. In the United States there is a large movement called *Cool Communities* that tries to do just that. *Cool Communities* combines government and non-governmental agencies in an attempt to mitigate the urban heat island effect. Light coloured surfaces and trees are two inexpensive ways to do this (McPherson 1994).

Most of the world's energy comes from burning fossil fuels or hydro-electric dams. Both are finite sources. Fossil fuels are becoming increasingly expensive in terms of explorations, removal, transport, and environmental damage. There is also a limit to the number of rivers that can be dammed for hydro-electricity. Cities require more energy per capita than rural areas and, as more people move into cities, the rate of electricity use increases disproportionately. We must discover alternative ways to use or save electricity. One of these ways is to reduce home and office energy consumption by moderating the microclimate around these spaces. Vegetative landscapes can reduce the energy required for residences by providing shade, wind control, and climate cooling (Parker 1983).

The energy savings from cooling are much greater than from heating, but both are substantial. Heisler (1986) found that it was possible to expect 10–12% savings on winter heating bills with windbreaks in place and a savings of up to 25% from shading in the summer. On average, trees in an optimal arrangement could save 20–25% of the annual energy bill for a house.

The arrangement and type of vegetation is very important because, in the summer, buildings are heated by the sun through windows, walls, and the roof (Akbari et al. 1989). Shading these areas, especially the windows, reduces heat gain (Heisler 1986). Tree arrangements that save energy provide shade for east and west walls and roofs (Heisler 1986, McPherson and Rowntree 1993). In all climates, a tree shading the west-facing wall of a wood frame building, exposed to the warmer afternoon sun, will provide twice the energy savings as the same tree on an east facing wall exposed to the cooler morning sun (McPherson and Rowntree 1993). It is also important to plant species that do not require a large amount of water, fertilizer or pesticides. Native trees and shrubs that are resistant to drought and disease and are appropriate to the site are best (Parker 1983).

Trees also remove heat from the air through evapotranspiration. A single tree can transpire 450 litres of water per day, consuming 1000 MJ of heat energy to drive the evaporation process (Bolund and Hunhammar 1999). When large trees are well distributed throughout a neighbourhood they can have a significant impact by lowering temperatures and reducing building energy use (Heisler 1986). In fact, large numbers of trees can reduce air temperature by 0.5–5 °C (McPherson 1994). Trees around parking lots can reduce air temperatures by 1–2°C more than untreed parking lots (Scott et al. 1999). This is important, as lower temperatures can lead to lower VOCs produced by starting automobiles, which in turn can lead to reduced ozone production. Parking lots are one area that can be easily targeted for urban cooling programs (McPherson 1989).

Simulations suggest that one 25ft (8.3 m) tree can reduce heating and cooling costs by 8–12% (McPherson and Rowntree 1993). Another study suggests that three trees, two on the west side of the building and one on the east, can reduce energy for cooling by 10–50% (Simpson and McPherson 1996). In Sacramento, annual savings of 12% were attributed to one 24 ft tree (Simpson 1998). Savings from additional trees was about 80% of the first tree (Simpson and McPherson 1996).

Evapotranspiration cooling effects can be 3–4 times greater than direct shade (McPherson and Rowntree 1993). This is especially evident when looking at the effects of turf grass. Having a lawn around a house (instead of asphalt or dirt) can lower house wall temperature by 6°C (Meier 1990–91) and provide cooling savings of 10–25% in hot climates like Arizona (McPherson and Rowntree 1993). In Tucson, Arizona the energy savings from lawns and shrubs were very similar but the maintenance costs for shrubs were much less. Turf requires more watering (McPherson et al. 1989). Trees, shrubs, or vines on the west wall of a home consistently lowered surface temperatures by 17% (Meier 1990–91).

In the winter the average house will exchange 75% of its air per hour, which causes one third of all heat loss (Heisler 1986). The proper placement and selection of species are crucial to lowering energy levels. On a regular 7,000ft^2 lot each tree will add almost 10% canopy cover. This will reduce wind speed by 5–15% and lead to an annual savings from wind shielding of 1–1.5% per tree (McPherson 1994). In areas with little cooling period, it is important to use solar friendly trees to make the most of that heat in winter (Simpson and McPherson 1996). The ideal tree is deciduous with a broad umbrella shaped crown, low to moderate water use, moderately pest resistant, and a low emitter of volatile organic compounds (Simpson and McPherson 1996).

Some requirements for winter wind protection and solar access in cold climates are (from Simpson and McPherson 1996):

- Dense evergreen foundation plantings
- Tall, dense evergreen/deciduous windbreaks, hedges and buffers
- Deciduous shade trees, shrubs, vines shading west walls and air conditioners (and in more temperate zones east walls)

- Unobstructed skyspace to the south for solar access
- Deciduous trees shading sidewalks, parking lots, and other paved surfaces
- Multistorey buffer plantings between neighbourhoods

In Chicago, on a per tree basis, annual heating can be reduced by 1.3%, cooling by 7%, and wind by 1–1.5% (McPherson 1994). In places where air conditioning is more prevalent such as Sacramento overall savings can be at least 12% of the annual energy bill (Simpson 1998).

Water

We depend on water for everything. Cities are especially problematic for completing the water cycle because most of the land is covered with impervious surfaces and water has no way of getting back into the ground. Vegetation can play a key role in helping to restore the water balance.

Built up area alter water flow with a high proportion of surface water run off. This increases the peak flood discharge, degrades water quality by picking up pollutants from streets, and causes groundwater depletion in cities because there is no replenishment (Bolund and Hunhammar 1999). In vegetated areas, only 5–15% of rainwater runs off the ground, while the rest is evaporated or infiltered. In cities with little vegetation, 60% of rain runs off and goes through storm drains (Bolund and Hunhammar 1999).

A row of narrow deciduous trees helps to keep a building cool in summer while allowing the sun through the glass in the winter

Each tree is projected to intercept 276 litres of storm water runoff annually (McPherson 1992). Storm water savings in Tucson were calculated at $0.18 per tree (Dwyer et al. 1992). In Chicago, avoided run off and water saved were valued at $3.75 per year per tree (McPherson 1994). In Modesto, street and park trees are estimated to reduce runoff by 292 million m^3, which is valued at $616,000 per year. Each tree reduces run off by 3.2 m^3 (845 gal) and is valued at $6.76 per year (McPherson et al. 1999). The savings may be low in Tucson because it is a desert and water is much more expensive than in other areas.

Trees intercept rainfall and reduce runoff. They function like retention/detention basins (temporary holding areas for water), which are essential to communities (Dwyer 1991). Storm water runoff from the parking lot can also be engineered to flow into planted areas, providing irrigation and reducing storm water runoff (McPherson 1989). It is important to try to reduce the amount of runoff going into the sewer system because it is expensive, less water is available for other uses, and runoff carries pollutants from the ground through the sewer system to the streams and oceans.

Broad leaf evergreens and conifers intercept more rainfall than deciduous trees because more rain falls in the winter when deciduous species have no leaves (McPherson et al. 1999).

Real Estate

Real estate prices also reflect ecosystem services. Real estate values are 7% more for homes on wooded lots than similar houses without trees (Dwyer 1991). In Columbus, Ohio, houses facing naturalized parks sold for an average of $1,130 more than similar houses one block away. Houses that faced recreational parks sold for $1,150 less. Open space parks contributed more to local property values than recreational parks (Dwyer 1991). This is likely because studies have shown that many people prefer forested park landscape to more open field-like appearance (Gobster 1991). Dwyer (1991) cites one court case where the loss of one tree was valued at 9% of the house value (in this case $15,000). A woman in north Vancouver, British Columbia was awarded $3,000 when her neighbour had several of her trees topped on two occasions (www.nsnews.com June 15, 2001)

It is reasonable to expect that in many instances property values would be higher in well-forested communities with a high percentage of tree cover. Urban forestry can be viewed as an investment that achieves an annual return in property taxes that are conservatively increased by 5% (Dwyer 1991).

Social/Health

The ecosystem provides many social and health benefits for us. For instance, hospital patients' recovery from surgery included a shorter hospital stay (by 10%), lower intake of narcotic pain drugs, and more favourable evaluation by nurses when their windows overlooked trees rather than a brick wall (Ulrich 1984). Exposure to urban forests for even brief periods of time can reduce stress and, therefore, improve personal health (Hull and Ulrich 1991). More trees should be planted where people can see them (workplace windows, roadways, hospitals, yards, other residential spaces). Trees save health care dollars by reducing hospital stays, doctor's visits, and absenteeism.

Willingness-to-pay surveys suggest that people are willing to pay $1.60 per visit more to see a site that was mostly wooded than a site with mowed grass with few trees (Dwyer et al. 1992).

Another reason for having nature in the city is that if people have to go less of a distance to enjoy nature, they will save money on fuel and reduce emissions. If just one gallon per year per individual were saved, that would add up to over $300 million US per year in avoided emissions costs (Dwyer et al. 1992).

Vegetation planting programs provide opportunities to teach inner city kids about nature, stewardship, and ownership of their neighbourhood (Hull and Ulrich 1991, Dwyer et al. 1992).

Noise is also a problem in cities. People prefer to walk down side streets that are quieter and have more vegetation than on busy throughways. Vegetation can reduce noise by 50% (Dwyer et al. 1992). Soft grass rather than concrete can reduce noise by 3 decibels and shrubs and trees can reduce it even more than that (Bolund and Hunhammar 1999). The cost of noise in the European Union is estimated to be between 0.2–2% of the GDP (Bolund and Hunhammar 1999).

The aesthetics of nature in the city can be psychologically restorative.

Kaplan and Kaplan (1995, 1998) present extensive evidence for the importance of nature in people's lives, especially their emotional and psychological well being. They have determined that the view from our window strongly influences how satisfied we are with our neighbourhood, that natural areas improve our self esteem and that green space is restorative to our minds and bodies. They conclude that many of the benefits attributed to a wilderness experience, far away from the city, can also be achieved in the city with attention to green space and biodiversity. The key components of the restorative experience provided by nature are the sense of being away, an awareness of other worlds, fascination and compatibility for action (for example, providing a quiet place to meditate).

Crime Prevention Through Environmental Design (CPTED), although a relatively new concept, is already an important initiative in many cities. The purpose of CPTED is to reduce crime by using design features that discourage criminal activity and encourage a healthy community. The concept includes principles of defensible space, surveillance, lighting, territoriality and landscaping. Vegetation figures prominently in such a design and this can add to biodiversity in the city.

There are concerns that some plantings may encourage crime by creating hiding places and hindering surveillance. However, the overall effect is positive when done properly. This is especially true if the use of greenspace creates an inviting atmosphere that increases the use of outdoor spaces by the community, combined with a sensitivity to sightlines for surveillance and perimeter plantings around properties for territoriality. The increased community presence discourages criminal activity (Kuo and Sullivan 2001). The sweet smells of flowers and vegetation may also make people happy and sociable; bad smells from garbage or car exhaust may make people tense and aggressive (Spears 2004).

Although it is clear that urban forests provide many invaluable ecosystem functions, it is hard to get a solid estimate on what they are worth. In the multi-year Chicago project, the annual financial benefits on one 30-year-old tree in a residential yard were valued at over US$300. If municipalities are going to invest further dollars in urban forestry, it would be worthwhile to consider an investment in yard trees over street trees. Yard trees are associated with lower planting costs, lower mortality rates, faster growth rates, and more effective building shade (McPherson 1992). In all cases, even Tucson, where water is expensive and scarce, it is still cost effective to have vegetation in the city rather than just pavement, rocks, and buildings. This chapter only deals with the costs that are relatively easy to measure. It is even more difficult to estimate the costs and benefits associated with increased biodiversity or improved quality of life.

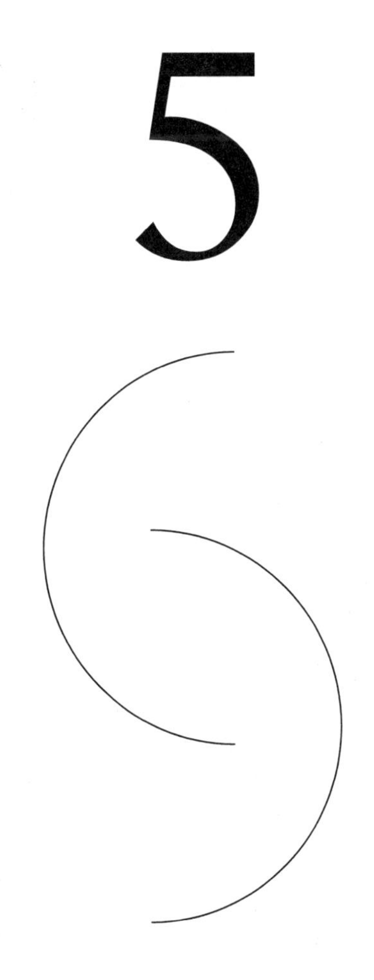

Valuating Nature's Services of Small-Scale Plantings in the City

Placing dollar values on the benefits of nature is not easy. It is much like asking, what is the value of a human body? Estimates of the value of a human body vary. Taken individually, a body's water (free), carbon (almost free) and other elements (usually cheap) can be had for as little as $20. They are, however, obviously worth more when they are considered together as a system that forms the human body. At the other extreme, some people may be insured for millions of dollars when the body is evaluated as a "life" that includes earning power and other factors.

Evaluating nature is no less complicated. It is not just an exercise in reducing it to its bare parts (like the $20 human body). It is also not as simple as looking at nature's earning power because evaluating socio-economic benefits is difficult — we do not know what they all are in many cases, and they may be impossible to measure accurately. We can make general estimates about the benefits of large areas, such as tropical rain forests, or an entire urban watershed. For example, the city of New York in 1997 signed a memorandum of agreement with communities in the Catskill/Delaware watershed and other agencies for the protection and compensation of rural areas whose drainage supplies the city with drinking water. The reason was to secure water that complied with the *Federal Safe Drinking Water Act*. Protecting the watershed enabled the city to obtain a waiver from the requirement that public surface water supplies must be filtered to minimize waterborne diseases, turbidity and phosphorus. Securing the natural filtration provided by the watershed prevented an expenditure estimated at $6 billion to construct a filtration facility and an annual expense of $300 million to operate it. Presented with these costs, can we assign a dollar value to one tree in the Catskill/Delaware watershed?

In another example, Oakville Green, a nonprofit organization dedicated to keeping Oakville, Ontario beautiful and healthy, has conducted assessments of the economic value of the city's natural heritage system. It includes an estimate that Oakville's 76,000 publicly owned trees have a collective economic benefit to the city of $70,000,000 (www.oakvillegreen.org February 2003). Does this mean that each tree is worth almost $1,000? Similar valuations for several American cities were done by American Forests using their CITYgreen program to analyse the information provided by the Geographic Information System (GIS) (www.americanforests.org, February 2003).

The New York City and Oakville, Ontario examples illustrate that, at the urban level or, even more difficult, the level of a backyard, valuating nature's services involves some art as well as science. What follows is an assessment that involves more science than art. The financial benefits presented are conservative. However, the amount by which they are increased largely depends on specific circumstances and personal values of the members of the communities in which the plantings occur.

Valuating Small Plantings

There are a number of estimates of the value of nature, but most are on large-scale benefits. For example, Whiting (2000) presents an estimated value for plants in Canada as $28.2 billion for both wild (forest trees, grazing, pharmaceuticals, etc.) and cultivated (forest plantations, horticulture, agriculture). Similarly, the combined value of wild and domesticated animals in Canada is 25.7 billion; all of Canada's biodiversity is valued at $70.6 billion.

Evaluating nature's services afforded by small scale plantings such as those found in yards is difficult because of the diversidty of species and their growth rates.

But what is the value of small plantings? How do they contribute to nature's services and what can be done to enhance these services in your backyard? Planting your backyard or balcony can save you money and improve the natural environment. It can provide a space for wildlife and increase the biodiversity and overall health of a region.

Calculating benefits of plantings usually addresses several key areas:

- energy savings for reduced heating in winter or cooling in summer
- particulates of less than 10 microns (PM_{10}) from automobile exhaust, industry and wood burning that cause lung disease
- ozone from the sun acting on smog from cars
- nitrogen dioxide from automobile exhaust that forms nitric acid
- sulphur dioxide, mainly from industry, that forms sulphuric acid

- carbon monoxide from automobile exhaust that binds with hemoglobin in red blood cells and reduces the oxygen carrying capacity of blood
- carbon dioxide from automobile exhaust that contributes to global warming
- hydrology, referring to reduced costs in removing storm water
- other benefits, primarily increases in property value from trees and landscaping

Tree size is often measured from the diameter of the trunk at chest height or about 1.3 m (4 ft) and is referred to as diameter at breast height (dbh).

The tables below can help you calculate what your yard is contributing to nature's services. Survey your yard. Add up how many square metres of lawn, shrub, and tree are present (be sure to calculate the canopy cover, i.e., the amount of actual ground space that the tree canopy takes up) and, if any are present, vines. Use the numbers from your yard and from Table 5.1 to calculate the total benefits of your yard. Compare this value to the amount of pollution produced by one person. Each person in the Greater Vancouver Regional District is responsible for 95 kg of carbon monoxide (CO), 24 kg of volatile organic compounds (VOCs), 21 kg of nitrous oxides (NOx), 2 kg of sulphur oxides (SOx), 8 kg of particulate matter, and 7,606 kg of greenhouse gases (www.gvrd.bc.ca, June 15, 2001).

McPherson et al. (2002) completed a detailed analysis of benefits and costs of urban trees in western Washington and Oregon. They calculated annual net benefits (total benefits minus planting and maintenance costs) for small-, medium-, and large-sized trees (28 ft tall, 25 ft spread; 38 ft tall, 31 ft spread; 46 ft tall, 41 ft spread, respectively) for both a public space and a private residential yard located opposite a west-facing wall 20 years after planting. Their results were:

	Leaf Surface Area	Private — West US$	Public US$
Small Tree	1,891 ft^2	$11.73	$ 5.22
Medium-sized tree	4,770 ft^2	$29.16	$23.30
Large Tree	6,911 ft^2	$51.46	$48.62

They also determined that a large tree adds $1,978 to the sale price of a home.

In 1998, residents of the GVRD added into the atmosphere: 191,000 tonnes of carbon monoxide, 48,000 tonnes of volatile organic compounds (VOCs), 42,000 tonnes of nitrogen oxides, 4,000 tonnes of sulphur oxides, and 16,000 tonnes of particulate matter Also, 15,334,000 tonnes of green house gases (such as carbon dioxide, ozone, and methane) were added to the atmosphere (www.gvrd.bc.ca, June 15, 2001).

Using the values from Table 5.4, it is possible to calculate that 3% of carbon dioxide, 15% of particulate matter, 2% of nitrogen oxide, and 50% of sulphuric oxides will be taken out of the atomosphere through plantings in residences. These values may seem insignificant, but they include residential land only.

Street trees, park trees, and natural vegetation also contribute to the amount of pollutants being removed from the air. For example, the 119,034 street trees in the city of Vancouver are estimated to be worth over $500 million in terms of their ecosystem services (Montpellier 2000).

Let's examine the contribution one backyard can make. Many of nature's services can be demonstrated in just one yard. A backyard that is planted with

Table 5.1 Air Pollution Reduction and Other Annual Benefits to the Environment for Different Vegetation Types

Energy	One Green Ash Tree (30 yr old — dbh of 30 cm)	One Square Metre of Shrub	One Square Metre of Grass	One Square Metre of Tree	One Square Metre of Vine
Cooling KWh savings	201.000	10.720	5.360	13.400	8.040
Heating Mbtu savings	8.300	0.443	0.221	0.553	0.332
PM_{10} (kg)	0.945	0.050	0.025	0.063	0.038
Ozone (kg)	0.297	0.016	0.008	0.020	0.012
Nitrogen dioxide (kg)	0.320	0.017	0.009	0.021	0.013
Sulphur dioxide (kg)	0.806	0.043	0.021	0.054	0.032
Carbon monoxide (kg)	0.072	0.004	0.002	0.005	0.003
Carbon dioxide (kg)	164.250	8.760	4.380	10.950	6.570
Hydrology (litre)	1125.630	60.034	30.017	75.042	45.025
Other benefits $US	233.820	12.470	6.235	15.588	9.353

Notes: The values for all of the tables have been determined using the values reported for the green ash tree (*Fraxinus pennsylvanica*) (McPherson 1994). The average green ash tree has a total leaf area of 77 m^2 and a leaf area index (LAI) of 5 (Nowak, 1994c), which means that the ground projection of the canopy of a green ash tree is approximately 15 m^2. By dividing the pollutant values for the green ash tree by 15 we get approximate values per metre square of ground projection. This can be used in turn to calculate the values for shrubs and grasses using their respective leaf area indices (LAI). For shrubs it is usually 4, and for turf grass it is usually 2 (Grack-Grzesikiewicz 1980). An approximation for vines is about 3.

For the pollutants, these values represent the amount taken in by the plant; for hydrology these values represent the amount of run off avoided and saved by plants; energy represents the amount of money saved on heating/cooling bills.

Prices used to estimate benefits (from McPherson 1994): PM_{10} — \$1.43/kg, CO_2 — \$0.0242/kg, ozone — \$0.539/kg, VOC — \$4.89/kg, SO_2 — \$1.804/kg, CO — \$1.012/kg, water avoided \$0.0758/litre, water saved \$0.0066/litre, other \$12.27 per cm dbh. Included in other benefits are scenic values, wildlife, improved water quality, noise abatement and social issues. Not included in these calculations are the savings from health care. One estimate for healthcare comes from the Lower Fraser Valley of British Columbia. Avoided emissions over the next 25 years will save C\$2.5 billion. Property values are also not included in these estimates. Property with backyard habitats (trees and shrubs) can increase property values by 5–9% of the house value. This can act as a revenue builder for municipalities as higher house values lead to higher property tax.

All prices are given in US dollars unless stated otherwise.

trees, shrubs, vines, and just a bit of grass will offer much more to the environment than one that is just made up of grass, cement and, perhaps, some annual flowers. The ideal backyard would be one that is full of many different plants, mostly native, has a driveway made of permeable materials such as crush rock (or no driveway at all), a rain barrel to capture roof run off, a composter, and a chair to enjoy the view.

By having less impermeable surfaces, run off is reduced. This allows water to percolate through the soil and the groundwater table to be replenished. It also reduces the amount of pollution going into the streams, as water running off pavement often carries pollution with it into the storm water drainage system. Erosion is avoided because water is not running over the land. There is

Table 5.2 Air Pollution Reduction and Other Annual Benefits to the Environment for Different Yard Types

Energy	One Yard All Grass	One Yard All Shrubs	One Yard All Trees	Ideal Yard	Realistic Yard
Cooling KWh savings	1072	2144	2680	2224.4	1474
Heating Mbtu savings	44.267	88.533	110.667	91.853	60.867
PM_{10} (kg)	5.04	10.08	12.6	10.458	6.93
Ozone (kg)	1.584	3.168	3.96	3.2868	2.178
Nitrogen dioxide (kg)	1.704	3.408	4.26	3.5358	2.343
Sulphur dioxide (kg)	4.296	8.592	10.74	8.9142	5.907
Carbon monoxide (kg)	0.384	0.768	0.96	0.7968	0.528
Carbon dioxide (kg)	876	1752	2190	1817.7	1204.5
Hydrology (litre)	6003.36	12006.72	15008.4	12456.972	8254.62
Other Benefits $US	1247.04	2494.08	3117.6	2587.608	1714.68

Notes: This table is based upon the average lot size 600 m2, of which 200 m2 is available for a wildlife habitat.

We hypothesize that the Ideal Yard is made up of 100 m^2 of trees, 50 m^2 of shrub, 50 m^2 of grass, and 10 m^2 of vines. This assumes no overlap of vegetation. The Realistic Yard is made up of 30 m^2 of trees (about 2 trees), 30 m^2 of shrub, and 140 m^2 of grass.

A realistic estimate is that the average residential lot has 200 m^2 of area available for wildlife habitat (Schaefer 1997). This is not necessarily indicative of all yards. Many houses have at least half of their yard available for wildlife habitat (~300 m^2). Table 5.2 gives some values for different yard types. This can be used as a rough guide for your own yard, if you don't want to measure it out, or as a comparison point for your yard.

The average residential green ash tree is worth about US$311 or about C$500 per year or US$21 per m^2 (about C$35) per year. Therefore, using the LAI once again, it is possible to determine values for other types of vegetation. Shrubs are worth $17 year/$m^2$, vines are worth $13 year/$m^2$, and grass is worth $8 year/$m^2$. The Ideal Yard is then worth US$3,480 (about C$5,600) year, the Realistic Yard is worth US$2,260 (about C$3,700) per year, the all grass yard US$1,600 (about C$2,600), the all shrub yard $3,400 (about $5,500), and the all tree yard $4,200 (about C$6,800).

more water available in the ground for plants to use rather than having to water them, which takes more water out of the reservoir. By having a rain barrel, it is easy to capture water coming off impermeable surfaces such as the roof and re-use it for watering plants.

Air filtration is another valuable service provided by backyards. Plants take in particulate matter, carbon monoxide, nitrogen oxides, sulphur oxides, and ozone. Cleaner air reduces human healthcare costs. Damage to plants, buildings, and cars is reduced when air is cleaner.

Plants around a house also save considerable money in energy costs. Conserving energy at the home can lead to reduced emissions from power plants. It can also reduce the need to build new power plants. This in turn reduces possible emissions and protects the environment as fewer rivers need to be dammed, less fossil fuel needs to be extracted from the Earth, and fewer trees need to be cut down for fuel.

By composting, the amount of waste going into the landfill is reduced. With a reduced need for landfills comes less wasted space and water contamination

Table 5.3 The Average Amount of Wildlife Habitat Available for Different Housing Types in the GVRD

Type of residence	Size of Yard m^2	Number in GVRD
Total		656,710
Single detached	200	292,040
Semi-detached house	93	14,755
Row house	23	45,315
Apartment, detached duplex	46	55,550
Apartment (5 or more stories)	30	75,055
Apartment (less than 5 stories)	12	170,145
Other single attached houses	23	1,145
Movable dwellings	1	2,695

Assumptions: For buildings greater than 5 stories we assumed an average of 6 units per floor, an average of 10 stories per building, with half of the units having balconies. It is also assumed that per balcony there is one m^2 of space available for wildlife habitat. Therefore there is 30 m^2 of habitat space per building greater than 5 stories.

For buildings less than 5 stories we assumed 6 units per floor, four stories, half with balconies, with one m^2 square of space for wildlife or 12 m^2 per building.

(Statistics Canada, 1996)

which often results from water leaching toxins from the garbage into the groundwater. Compost can also be used as a soil conditioner in the garden, leading to healthier soil. Healthy soil leads to increased biodiversity of insects and other creatures that live in the soil. This feeds the next trophic level and enhances the food web.

Plants also sequester carbon dioxide. Carbon dioxide is the major component of greenhouse gases and the leading cause of the global warming trend. Global warming could have catastrophic effects on the survival of many species of plants and animals. By removing carbon dioxide from the atmosphere, the effects of global warming will be slowed down.

There are reduced health expenses associated with this type of yard. Health benefits and reduced health care costs result from cleaner air, exercise associated with gardening, and reduced stress by having a sanctuary to enjoy.

By having more plants in a yard, more wildlife will be attracted to it. Pollination is one of the most important of nature's services. Pollinators increase biodiversity by ensuring that plants reproduce and provide food for people and wildlife. Heavily planted backyards increase the amount of habitat available for wildlife. This will allow more wildlife to live in the city, increasing the biodiversity and overall health of the ecosystem. This space can also act as a corridor of safe travel to larger green spaces.

Table 5.4 Air Pollution Reduction and Other Annual Benefits to the Environment for Different Housing Types Throughout the Greater Vancouver Regional District

Energy	Single Detached	Semi-Detached	Row House	Apartment Detached Duplex	Apartment (5+ stories)	Apartment (less than 5 stories)	Other Single Attached House	Movable Dwelling	Total
Cooling KWh savings	430,466,960	10,113,224	7,681,345	18,832,561	1,659,660	15,047,623	194,088	19,862	498,950,326
Heating Mbtu savings	17,775,501	417,610	317,189	777,662	685,252	621,369	8,014	820	20,603,421
PM_{10} (kg)	2,023,837	47,547	36,113	88,541	78,019	70,746	912	93	2,345,811
Ozone (kg)	636,063	14,943	11,350	27,827	24,520	22,234	286	29	737,254
Nitrogen dioxide (kg)	684,249	16,075	12,209	29,935	26,378	23,918	308	31	793,107
Sulfur dioxide (kg)	1,725,080	40,528	30,782	75,470	66,502	60,302	777	79	1,999,524
Carbon monoxide (kg)	154,197	3,622	2,751	6,745	5,944	5,390	69	7	178,728
Carbon dioxide (kg)	351,762,180	8,264,164	6,276,920	15,389,294	13,560,562	12,296,379	158,602	16,230	407,724,334
Hydrology (litre)	2,410,679,224	56,635,566	43,016,682	105,465,152	92,932,575	84,268,939	1,086,927	111,231	2,794,196,299
Other Benefits $US	68,284,792	1,604,256	1,218,488	2,87,401	2,632,404	2,386,998	30,788	3,150	79,148,280

Notes: Table 5.4 uses the values from Table 5.2 for the Realistic Yard and the amount of space available according to Table 5.3 to determine the effects all of the yards in the GVRD have on pollution.

Table 5.5 Air Pollution Reduction and Other Annual Benefits to the Environment for Different Vegetation Types on Boulevards

Energy	One Boulevard — All Grass	One Boulevard — All Shrub	One Boulevard — All Tree
Cooling KWh savings	2444.160	4888.320	6110.400
Heating Mbtu savings	100.928	201.856	252.320
PM_{10} (kg)	11.491	22.982	28.728
Ozone (kg)	3.612	7.223	9.029
Nitrogen dioxide (kg)	3.885	7.770	9.713
Sulphur dioxide (kg)	9.795	19.590	24.487
Carbon monoxide (kg)	0.876	1.751	2.189
Carbon dioxide (kg)	1997.280	3994.560	4993.200
Hydrology (litre)	13687.661	27375.322	34219.152
Other Benefits $US	2843.251	5686.502	7108.128

Notes: The average boulevard is 3 m by 152 m or 456 m^2.

The values that can be attained from planting boulevards with different vegetation types are presented in Table 5.5.

An all grass boulevard would be worth US$3,648 (about C$6,000) per year, all shrubs US$7,752 (about C$12,600) per year, and all trees $9576US (about C$15,500).

Table 5.6 Air Pollution Reduction and Other Annual Benefits to the Environment for City Blocks with Different Vegetation Types

Energy	One Block — Grass	One Block — Shrub	One Block — Tree	One Block — Realistic
Cooling KWh savings	21440.000	42880.000	53600.000	29480.000
Heating Mbtu savings	885.333	1770.667	2213.333	1217.333
PM_{10} (kg)	100.800	201.600	252.000	138.600
Ozone (kg)	31.680	63.360	79.200	43.560
Nitrogen dioxide (kg)	34.080	68.160	85.200	46.860
Sulphur dioxide (kg)	85.920	171.840	214.800	118.140
Carbon monoxide (kg)	7.680	15.360	19.200	10.560
Carbon dioxide (kg)	17520.000	35040.000	43800.000	24090.000
Hydrology (litre)	120067.200	240134.400	300168.000	165092.400
Other Benefits $US	24940.800	49881.600	62352.000	34293.600

Notes: The values that can be attributed to one average city block are presented in Table 5.6. These numbers were determined by multiplying the values for the Realistic House by the number of houses found on the average city block (about 20).

Table 5.7 Air Pollution Reduction and Other Annual Benefits to the Environment Based on Different Tree Species

Energy	One Green Ash Tree — LAI of 5	Trees with LAI of 8	Trees with LAI of Approx. 4.5	Trees with LAI of Approx. 4	Trees with LAI of Approxi. 3.5
Cooling KWh savings	201.000	321.6	180.9	160.8	140.7
Heating Mbtu savings	8.300	13.28	7.47	6.64	5.81
PM_{10} (kg)	0.945	1.512	0.8505	0.756	0.6615
Ozone (kg)	0.297	0.4752	0.2673	0.2376	0.2079
Nitrogen dioxide (kg)	0.320	0.5112	0.28755	0.2556	0.22365
Sulphur dioxide (kg)	0.806	1.2888	0.72495	0.6444	0.56385
Carbon monoxide (kg)	0.072	0.1152	0.0648	0.0576	0.0504
Carbon dioxide (kg)	164.250	262.8	147.825	131.4	114.975
Hydrology (litre)	1125.630	1801.008	1013.067	900.504	787.941
Other Benefits $US	233.800	374.08	210.42	187.04	163.66

Notes: Green ash have a leaf area index (LAI) of 5. Lombardy poplar (*Populus nigra*) has an LAI of 8. Urban trees with an approximate LAI of 4.5 are Norway maple, little leaf linden, European mountain ash (*Pyrus aucuparia*), and horse chestnut (*Aesculus hippocastanum*). Black locust, silver maple (Acer saccharinum), lime linden (*Tilia sp.*), English oak (*Quercus robur*), sycamore (*Platanus occidentalis*) are trees having an approximate LAI of 4. Common ash, *Malus spp.*, and white willow are examples of species with an LAI of approximately 3.

Instead of having just grass, gravel or landscape blocks, a backyard could be turned into a wonderful wildlife habitat to increase biodiversity in the region.

Yards

Let's examine two identical yards right beside each other. Yard A is made up mostly of grass, with one small tree in the backyard (Table 5.8). Yard B is a wildlife haven: it has several large trees, shrubs everywhere, a small amount of grass, and vines growing on the back wall (Table 5.9).

Table 5.8 Yard A

Energy	Tree	Grass	Total
Cooling KWh savings	201	991.6	1192.6
Heating Mbtu savings	8.3	40.94667	49.24667
PM_{10} (kg)	0.945	4.662	5.607
Ozone (kg)	0.297	1.4652	1.7622
Nitrogen dioxide (kg)	0.3195	1.5762	1.8957
Sulphur dioxide (kg)	0.8055	3.9738	4.7793
Carbon monoxide (kg)	0.072	0.3552	0.4272
Carbon dioxide (kg)	164.25	810.3	974.55
Hydrology (litre)	1125.63	5553.108	6678.738
Other benefits $US	233.82	1153.512	1387.332

The yard is 200 m^2 of which there is one tree (15 m^2) and 185 m^2 of grass. Yard A is worth US$1795 (about C$3000) annually.

Table 5.9 Yard B

Energy	Trees	Shrubs	Grass	Vines	Total
Cooling KWh savings	1005	1018.4	160.8	80.400	2264.600
Heating Mbtu savings	41.5	42.05333	6.64	3.320	93.513
PM_{10} (kg)	4.725	4.788	0.756	0.378	10.647
Ozone (kg)	1.485	1.5048	0.2376	0.119	3.346
Nitrogen dioxide (kg)	1.5975	1.6188	0.2556	0.128	3.600
Sulphur dioxide (kg)	4.0275	4.0812	0.6444	0.322	9.075
Carbon monoxide (kg)	0.36	0.3648	0.0576	0.029	0.811
Carbon dioxide (kg)	821.25	832.2	131.4	65.700	1850.550
Hydrology (litre)	5628.15	5703.192	900.504	450.252	12682.098
Other benefits $US	1169.1	1184.688	187.056	93.528	2634.372

The yard is made up of 75 m^2 of tree, 95 m2 of shrub, 30 m^2 of grass, and 10 m^2 of vine. Yard B is worth US$3560 (about C$5800) annually. These values do not include healthcare savings or real-estate benefits. Yard B has a substantial increase in monetary worth over Yard A. It also helps protect the environment, and provides more habitat for wildlife and greater opportunities to enjoy nature.

Balconies

The average apartment balcony is approximately 2 m by 4 m, or 8 m^2. There is also vertical space. Plants could be hanging from the ceiling or attached to railings. Many of nature's services that apply to backyards also apply to balconies.

Table 5.10 Environmental Benefits of Balcony Habitat

Energy	Shrubs	Vines	Perennials	Total
Cooling KWh savings	21.440	16.080	32.160	69.680
Heating Mbtu savings	0.885	0.664	1.328	2.877
PM_{10} (kg)	0.101	0.076	0.151	0.328
Ozone (kg)	0.032	0.024	0.048	0.103
Nitrogen dioxide (kg)	0.034	0.026	0.051	0.111
Sulfur dioxide (kg)	0.086	0.064	0.129	0.279
Carbon monoxide (kg)	0.008	0.006	0.012	0.025
Carbon dioxide (kg)	17.520	13.140	26.280	56.940

The difference between the balcony above with no plants and the one below is dramatic. Imagine entire apartment buildings with gardens on their balconies and the amount of habitat that creates.

Planted balconies will filter air, cool buildings, and provide habitats for wildlife. Birds have been known to nest in simple terra cotta pots on the 24th floor of a downtown apartment building. This space is a valuable resource. If you think of a tall building it may seem barren, but if each of those balconies had just a few plants, the building is transformed into an urban oasis where a hummingbird can get a quick meal or a songbird can make its home.

The balcony from Table 5.10 is made up of 2 m^2 of shrubs, 4 m^2 of perennials, and 2 m^2 of vines. It is assumed that perennials have the same impact as vines.

The vegetation that made up this balcony:

- Western trumpet honeysuckle along one wall (2 m^2).
- Two large pots with shrubs, one red-flowering currant and one snowberry. Each plant is about 1 m^2.
- Three hanging baskets (1/2 m^2), one large pot (1 m^2), and three railing planters (1/2 m^2) of wild flowers, perennials, and ferns.

These values may seem small, but they can be compared to blue box recycling efforts. Each person's contribution may not seem significant, but the collective effort is substantial. Since 1990, in the GVRD recycling has led to a 38% decrease in the amount of material going into the landfill. This is all the result of those small blue bins. The same logic can work for small spaces. If each person adds a few plants, a tremendous difference can be made.

Municipalities

There are also many things that can be done on a municipal level with regard to public spaces:

- Pesticide and fertilizer use can be reduced or not used at all
- Grass can be removed and replaced with wildflower beds and shrubs
- Naturescape principles can be employed (native plants adapted to the area)
- Trees can be planted near parking lots to cool those areas, thereby reducing ozone formation and air pollution
- By-laws can be passed to restrict trees being cut on private property and the amount of trees being planted in new developments

Examples of Municipal Initiatives

The city of Coquitlam, BC and Douglas College added an arboretum to city property adjacent to the College. This area was previously a grassy field. In the spring of 2000, 66 trees of approximately 3 m^2 each of canopy cover were planted. In 20 years these trees will have grown to at least 15 m^2 of canopy cover each. (*See* Table 5.11 for the difference these trees made when just planted to the difference they will make in 20 years.)

The city of Port Moody, BC is another example of a municipality that is trying to make a difference by setting an example. The city has instituted Naturescaping principles into its own maintenance practices. They are no longer using pesticides or herbicides to remove insects or weeds. They will only remove trees that are a hazard, not because they are blocking a view or creating shade and they are trying to use native plants in their street and park plantings. This will create much-needed habitats in urban environments. They have also planted demonstration gardens for residents to view native plants and what they can do in their yards. They are also encouraging residents to use these principles in their own yards.

Nature's services are also being thought of on a larger community scale. The East Clayton project is a 250 hectare (13,000 people) housing (community) project in Surrey, BC that attempts to fit into the landscape rather than the other way around. When built, it will have its own storm water management system of ponds that will not be emptied into the local streams. It will have one-third less blacktop than a regular suburb. There will be a reduction of 40% or greater in vehicle miles traveled per day, which equals a reduction of greenhouse gas emissions of 40%. There will be nature in the community, smaller streets with trees, parking lots with trees, and services and jobs within the community. Their mandate includes four things: affordable housing, accessible tran-

Table 5.11 Environmental Benefits of Street Trees

Energy	One m^2 of Tree	198 m^2 of Tree (66 Trees by 3 m^2) 2000	990 m^2 of Tree (66 Trees by 15 m^2) 2020
Cooling KWh savings	13.400	2653.2	13266
Heating Mbtu savings	0.553	109.56	547.8
PM_{10} (kg)	0.063	12.474	62.37
Ozone (kg)	0.020	3.9204	19.602
Nitrogen dioxide (kg)	0.021	4.2174	21.087
Sulphur dioxide (kg)	0.054	10.6326	53.163
Carbon monoxide (kg)	0.005	0.9504	4.752
Carbon dioxide (kg)	10.950	2168.1	10840.5
Hydrology (litre)	75.042	14858.32	74291.58
Other benefits $US	15.588	3086.424	15432.12

sit, commercial services and jobs within the community, and natural systems preserved and enhanced.

Some of the goals for tree cover when mature:

- Parking lots — 50%
- Residential Yards — 40%
- Streets — 60%
- School/Parks — 50%
- Greenways — 50%
- Boulevards — 60%

These numbers are quite remarkable goals since most cities have about 10% tree cover. New suburbs usually have quite a bit less than this.

Everything Matters

There is a perception that wildlife habitat only exists in parks and other protected areas. Also, the only habitat worth creating is one that would support fish or breeding bird populations. However, small plantings add up. The yards in Greater Vancouver, for example, collectively can create habitat equal to six Stanley Parks.

Even if the habitat in yards and balconies is not of great quality, it can help to buffer the parks and protected areas from urban impacts. It can provide a safety net that wildlife can use during critical times when food is in short supply in the parks, or weather is bad. Biodiversity conservation in cities needs to look beyond the major green spaces and include backyard and balcony habitat. Everything matters!

Part 2

Urban Habitats

Forests
Shrub Communities
Freshwater Landscapes
Open Spaces
Barren Landscapes
Paved Landscapes
Corridors
Garden Landscapes
Public Landscapes

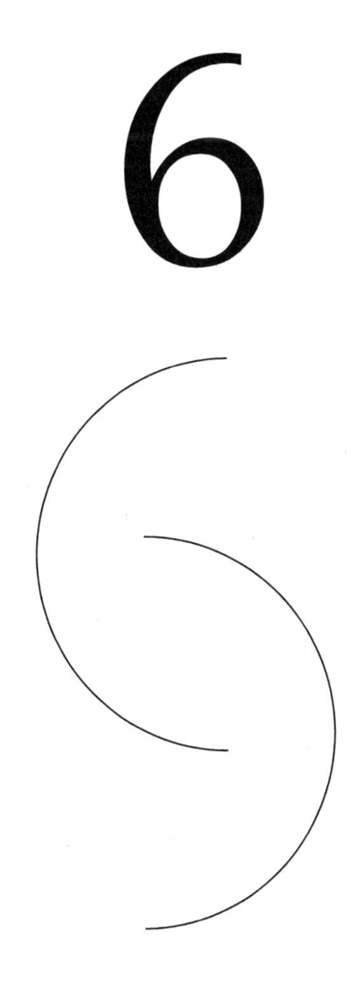

Forests

At the city centre, people are more abundant and there is much less vegetation. This is the urban landscape gradient, with people increasing and plants decreasing the closer you get to the downtown core. Species diversity also typically decreases and there may be less understorey vegetation for safety and aesthetic reasons. Also following the urban landscape gradient is a gradient of discrete habitat types, such as patches of forest, versus diffuse habitat such as backyards and street trees.

Urban forests include street trees, park trees, green spaces, residential land, and public and private spaces with vegetation (Bradley 1995). They provide unique habitat islands and can encourage, rather than displace, wildlife. They daily remind the public of the ecological processes that we depend on for survival (Whitney 1987).

Urban forests provide cities with scenery, buffers, wildlife, and energy and water conservation (Bradley 1995). The size and species composition of urban forests change with proximity to city centres. Forest fragments on the edge of urban areas are usually larger, with greater species diversity and more understorey vegetation.

Urban forests differ from their rural counterparts. Trees in urban centres have a much shorter lifespan than elsewhere (Figure 6.1). Urban trees are subjected to greater stress such as pollution, drought, and flood (Garber 1987). Fragments of forest surrounded by a city are not representative of the original stand. The harsh environment of the city is largely responsible for this change. Trees in urban centres have lower survival rates than rural trees with longevity decreasing as proximity to the city core increases. Reduced vigour is likely due to factors such as higher toxin concentrations from vehicles and industrial waste, dumping, and vandalism, including arson and tree cutting.

Trees have a major influence on urban biodiversity through their canopy cover. Shade from trees over streets is effective in mitigating urban heat islands (Akbari et al. 1992) and can extend the lifespan of pavement (McPherson et al. 1999). Trees can improve air quality, lowering ground level ozone by reducing hydrocarbons in the air from automobile exhaust (Scott et al. 1999). They also help to control stormwater runoff (Xiao et al. 1998).

The amount of canopy cover over streets and sidewalks in a city varies by zone, with one study measuring 4–46% canopy cover in Davis, California, with a citywide average of 14% (Maco and McPherson 2002). The study concluded that a canopy coverage over streets and sidewalks of 25% is attainable considering land use, planting sites and age distribution. However, establishing targets for optimal canopy coverage is problematic because of difficulties in measurement and correlating canopy cover with environmental benefits.

Mature Forest Fragments

Fragments of second, third or old growth forests are rare in urban areas. They are often either left intact during the development of an area or are located in areas that are not suitable for development, such as ravines.

The quality of urban forests as wildlife habitat is extremely variable. The best may contain intact stands of mature trees with a well-developed understorey and have biodiversity similar to that of the wilderness fringe on the outskirts of the city. Such forests are valued as wildlife habitat and their recreational value to the community may be limited to a few trails. Their ecological value is great because of the biodiversity of plant and animal life they support. They also help maintain water quality in the creeks within the watershed and are an important buffer to noise.

Forest fragments may instead be more intensely managed, and may or may not have a defined understorey. In some urban areas, understorey vegetation is

Figure 6.1 Average Tree Lifespan

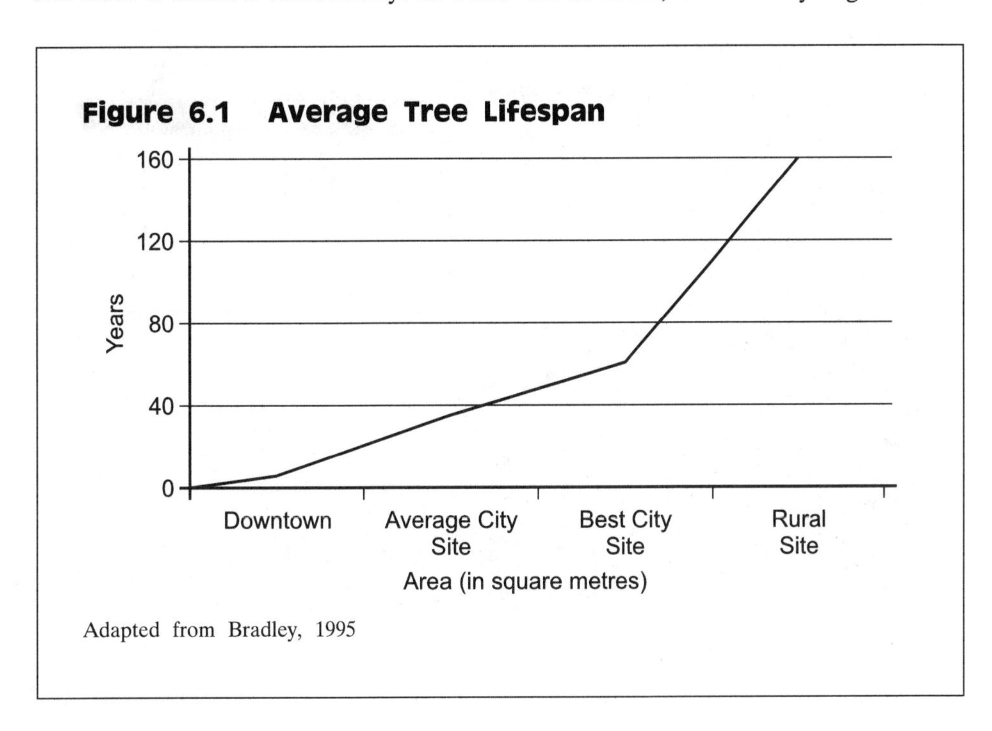

Adapted from Bradley, 1995

removed for safety reasons. Tree species in mature forest fragments vary with location, from coniferous to deciduous stands or a combination of both.

Mature forest patches are continually going through ecological succession. Although these patches would be expected to eventually reach a mature climax/ old growth community, in urban areas, forest fragments usually disturbed by development do not reach their climax (Bradley 1995).

Mature forest fragments have a significant amount of biomass in their leaves and on the ground. This supports insect larvae, small mammals, birds and other wildlife. Snags often associated with mature forests offer nesting habitat for woodpeckers, swallows and chickadees (Link 1999).

Forests with little or no understorey vegetation provide no cover and limited food sources for most animals, except for species such as the gray squirrel (*Sciurus carolinensis*) and Douglas squirrel (*Tamiasciurus douglasii*). The forest soil is also low in invertebrate number and diversity; the floor contains little or no leaf litter and, in coniferous stands, the soil remains acidic and dry.

In urban forests where the understorey has been removed, species diversity is naturally limited to the trees that remain. Such areas display a forest floor nearly devoid of vegetation except for species such as horsetail, which is capable of inhabiting harsh environments, and a few ferns, where gaps in the canopy have permitted some sunlight to pass through.

Mature forest fragments enjoy a high diversity of wildlife by providing food, shelter and nesting sites. The urban forest comprises a mix of coniferous and deciduous species with a well-developed understorey allowing for such diversity. These mixed forests also enable many animals to remain in the area over the

Snags in forests provide important habitat for a wide range of wildlife.

winter because conifers and evergreen shrubs provide good cover and nesting sites for birds, rodents and larger mammals such as coyotes and deer.

The size of habitats and their proximity to the city are two factors contributing to faunal number and diversity. A small urban forest close to the centre of a city limits diversity: food sources are not as plentiful and other factors, such as increased environmental disturbance, come into play. For example, in a small closed-canopy urban forest, common city birds such as Crows (*Corvus brachyrhynchos*) and Black-capped Chickadees (*Parus atricapillus*) will likely outnumber more intolerant species such as Swainson's Thrush (*Catharus ustulatus*) and Steller's Jay (*Cyanocitta stelleri*).

Woodlots

Urban woodlots vary in size, age, and species composition due in large part to differences in their size, past use, present management, and disturbance. The majority of urban woodlots are formed through spontaneous natural regeneration on abandoned lots and old agricultural land.

A woodlot is usually large enough to have many tall, often thin-boled trees growing close enough together to form a canopy. The canopy regulates the amount of light reaching the understorey, thereby influencing the composition of the understorey vegetation growing. Other significant factors affecting the understorey are the age of the woodlot, soil conditions, the availability of water, and successional status. In turn, successional status is a function of past use, present management, and disturbance.

A typical woodlot consists of a number of closely spaced, fast growing, pioneer species such as alder (*Alnus sp.*), cottonwood (*Populus sp.*), and willow (*Salix sp.*), that are capable of forming a canopy. Woodlots composed of mixed tree species usually do not have a single distinct canopy layer, but may display different levels of cover consisting of the main canopy, a sub-canopy of understorey trees, shrubs, and wildflowers. As the layers of vegetation in a woodlot increase, the complexity of the site also increases, allowing the woodlot to support a greater biodiversity.

Woodlots can be partially or totally deciduous depending on climate, soil, and past and present use. Alder, for example, pioneers on sites with poor mineral soils and neutral pH. Birch (*Betula sp.*) favours damp sites with more acidic soils, while cottonwood is dominant on wetter sites. Scouler's willow (*Salix scouleriana*) is tolerant of drier conditions and is capable of pioneering sites with little soil.

The dominant understorey species, if present, are fast colonizing shade-tolerant shrubs and wildflowers. If the site is near water, ferns and other species tolerant of wet sites will also thrive. Seedlings and saplings are usually not the same species as the taller fast colonizing species that form the canopy. However, they make up the next wave of secondary growth trees, which if left undisturbed, will replace the primary colonizers through ecological succession.

As the complexity and species diversity of urban woodlots increase, they are able to support more birds. Bird species diversity often correlates with foliage height complexity (MacArthur and MacArthur 1961). When forests meet a different habitat type, such as grassland, vegetation and wildlife diversity increase at the edges.

Urban woodlots with closely spaced trees and large amounts of understorey growth provide excellent habitat for a variety of wildlife. The presence of an extensive shrub layer, foliage height complexity, and freedom from disturbance can greatly increase the diversity of bird life in woodlots (MacArthur and MacArthur 1961, Gilbert 1991). Larger and more productive woodlots usually accommodate more species. High productivity and biomass translate into more space for more species (Janzen 1976). Both density and number of species of mammals correlate positively with soil fertility and undergrowth density.

Woodlots offer shelter for wildlife, taking advantage of food from adjoining homes and businesses. Raccoons (*Procyon lotor*) and crows are two examples of species that leave woodlots to forage for food in surrounding residential areas.

Microclimatic conditions, especially if the leaf litter and understorey remain, allow for a relatively high degree of biodiversity. Typically, in an urban forest some sections of the canopy are not representative of true climax species. This layering effect is the result of different seral stages of succession. For example, faster growing alder or vine maple (*Acer circinatum*) may succeed young conifers. There may be a mix of shade intolerant species, such as pines (*Pinus sp.*) and larch (*Larix sp.*) with shade tolerant species, such as the western hemlock and western red cedar.

Woodlots with their extensive edge effect offer abundant biodiversity.

Subcanopy tree layers are typically composed of species such as Pacific dogwood (*Cornus nutallii*), salmonberry (*Rubus spectabilis*), trailing blackberry (*Rubus ursinus*), and dull Oregon grape (*Mahonia nervosa*). Herbaceous species of the forest floor include ferns, a variety of mosses, and flowering plants, such as Pacific bleeding heart (*Dicentra formosa*), false lily-of-the-valley (*Maianthemum dilatatum*), and western buttercup (*Ranunculus occidentalis*).

Linear Forests

Linear forests are elongated and are usually a uniform width. They can form closed loops, as in the case where they may encircle a small urban island or a large suburban estate. More frequently, however, they generally follow a linear, winding route, particularly if they are associated with natural or artificial features such as roads, waterways, railways, and utility rights-of-way. In urban centres located on islands or adjacent to large bodies of water, they may constitute an integral part of a continuous linear park system adjacent to the shoreline. In other parts of the urban landscape, they may be found on the edges or slopes of natural features such as ravines or bluffs.

The same mechanisms that govern understorey growth, succession, and species diversity in woodlots are also at work in these linear habitats. Since the main difference between the woodlot and the linear forest is, essentially, the shape of the habitat, similar general patterns emerge with differences due primarily to its linear nature.

Compared to forest fragments and woodlots, linear forests have less leaf litter. Soil development and characteristics are determined by surrounding land uses.

Because of the potential of a linear forest to traverse a number of distinct topographical regions, the aggregate tree and plant species characteristic of each of these individual sectors may be able to accommodate and support a greater diversity of wildlife than a woodlot habitat located on a more homogenous site. The potentially greater structural complexity provides more ecological niches for wildlife to exploit. The diversity of terrain, soils, and moisture regimes primarily influences the types of flora that occurs in these distinct regions, and in turn the diversity of the fauna.

Another distinct difference between linear forests and woodlots is that the linearity provides a somewhat less restricted or confining environment for wildlife species needing range habitat. If the linear habitat is sufficiently extensive and connects other widely spaced urban habitats, it becomes even more useful and important for those species as a natural corridor through which they can travel and forage. This is particularly true for linear forests bordering waterways where the forests make the foreshore area less accessible to human traffic.

Interruptions and discontinuities in a linear forest caused by roads or highways are significant hazards for wildlife using linear habitats as corridors because of automobiles traffic.

Linear forest may have a high biodiversity due to edge effects. In general, some of the most conspicuous and commonly occurring wildlife includes abundant bird communities that make use of any available cover, nesting areas, perching sites, and food supply. Nesting crows, foraging raptors along the forest edges, and feeding flocks of Black-capped Chickadees, are common examples. Sections of a linear forest that have moderate to heavy understorey growth and are adjacent to a waterway can also provide cover and protection from disturbance to many species using the foreshore areas. Examples include Mute Swan (*Cygnus olor*), Great Blue Heron (*Ardea herodias*), and a variety of ducks and geese.

Other wildlife inhabiting the linear forests, such as raccoons and coyotes, may not restrict their foraging to these areas, but may venture abroad into any adjacent urban development.

Nature's Services Provided by Urban Forests

Some of the benefits of urban forests have been described earlier in Chapter 4 as part of the general discussion of urban ecosystem services. Presented here is another overview with a few additional considerations.

Urban forests provide many benefits both to urban ecosystems and residents. These include the reduction of urban heat island effects, the presence of wildlife in neighbourhoods, and a substantial contribution to the physical, biological, and psychological well being of individuals and communities. Urban forests also provide direct access to recreation, wildlife, scenery, and a variety of other amenities (Bradley 1995).

Forest ecosystems are one of the most important ecosystems on this planet. Landscapes are stabilized and soil erosion is prevented through forest root systems and above ground vegetation (Daily 1997). By protecting soils, urban forests help retain moisture, as well as store and cycle nutrients through soil and ecosystems (Daily 1997). Urban forests preserve watershed function by regulating water flow (both quantity and quality). Forests also moderate flood and drought regimes, as water is stored as groundwater. Street trees and urban forests reduce the urban heat island effect by reducing the local temperature (Bradley 1995). Carbon dioxide is sequestered by trees, thereby containing global warming to some extent (Daily 1997). Forests also remove harmful chemicals from the air.

Forests regulate climate on a micro scale, reduce noise, act as refugia for wildlife, provide protection from the sun's ultraviolet rays, and add to the aesthetic beauty of the surrounding landscape (Constanza et al. 1997).

The following are a few of nature's services provided by forests as presented in Daily (1997) (Note: All monetary references from Daily from here on in, unless specified otherwise, are in US funds):

- Stabilize landscapes
- Protect soils and store nutrients
- Buffer against the spread of pests and disease
- Regulate quantity and quality of water flows
- Help prevent flood/drought regimes in downstream territories
- Critical to energy balance of Earth
- Help to contain global warming
- Excess siltation reduces productivity and operation life of dams
- Meeting basic needs for water and sanitation for entire developing world would cost $300 billion over a 10-year period; even if watersheds reduced this need, only 1% still worth $3 billion
- Sun's radiational heating of the Earth's surface (albedo effect) depends on vegetation, which absorbs more heat than bare soil; moisture from evapotranspiration condenses to form rain
- Forests hold 1,200 gigatonnes of carbon in plants and soil of a total of 2,000 gigatonnes for all terrestrial plants and soil
- Of carbon in forests, 50% in boreal (Earth's largest terrestrial biome), 33% in tropical; 66% is contained in soils and peat deposits

- Boreal forest has more carbon than all known fossil fuel reserves
- Forests account for 65% of all net plant growth and carbon fixation on land
- Of 7.6 gigatonnes of carbon emitted per year into global atmosphere, 1.6 gigatonnes comes from burning in the tropics
- Acid precipitation in northern Europe resulting in commercial losses to the forest industry of $30 billion per year
- Tropical rain forests have 50% of all species on Earth.

7

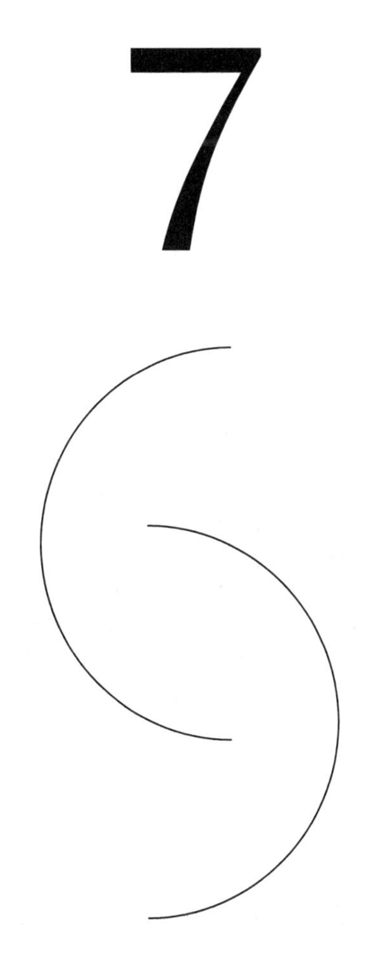

Shrub Communities

Shrubs are woody plants that are usually less than 10 m tall at mature height (Pojar and MacKinnon 1994). They are also usually multi-stemmed, whereas trees usually have only one main stem. Shrub communities occur when shrubs are the dominant or only plants present. When shrubs dominate a site, they often grow very densely and few other species can grow in the understorey. Shrub communities can be found in association with many other urban habitats, including grasslands, forests, ponds, and roadsides. There are many different types of shrub communities. Thickets and hedgerows will be discussed in this chapter.

Thickets

Thickets are areas where shrubs form a thick mass that cannot be seen through. Typically, only the outer edge of the thicket is visible. They range in size and can be as small as one square metre or as large as several hectares. Thickets tend to be largely homogenous with one dominant species, such as hardhack (*Spiraea douglasii*), red-osier dogwood (*Cornus stolonifera*), red elderberry (*Sambucus racemosa*), salmonberry (*Rubus spectabilis*), thimbleberry (*Rubus parviflorus*), Scotch broom (*Cytisus scoparius*), or salal (*Gaultheria shallon*). They provide cover for wildlife to feed and breed. Thickets growing in the forest understorey add another dimension to the ecosystem and thereby increase biodiversity.

Some thickets, especially those made up of woody shrubs, like hardhack, are able to reach an almost climax-like dominance on some sites. The denseness of their growth and the shading caused under the canopy restrict most other plants from growing in the understorey.

Wildlife that find thickets attractive include black bear (*Ursus americanus*), Savannah Sparrow (*Passerculus sandwichensis*), Ring-necked Pheasant (*Phasianus colchicus*), coyote (*Canis latrans*), and voles (*Microtus sp.*). Some thickets provide berries for a variety of songbirds while others have tender shoots and leaves that voles and deer can browse. Ring-necked Pheasants and Savannah Sparrows are able to find protected nest sites deep in thickets, while coyotes may be able to create a den by pushing through the outer layer of growth to the empty interior of some shrubs. Thickets are important in urban areas because they provide important shelter for many wildlife to nest, den, and raise young.

Blackberry Thickets

Blackberries are shrubs with woody, persistent stems (canes) and no distinct trunk, and are capable of reaching heights of up to three metres. The upright or arched branching canes usually form dense thickets along roadways, edge habitats, or in forest understorey. Blackberry stems are well armed with strong sharp prickles and are capable of growing so densely the thickets are often impenetrable.

The four species of blackberry commonly found in urban habitats are the trailing or wild blackberry (*Rubus ursinus*), sub-alpine blackberry (*Rubus nivalis*), Himalayan blackberry (*Rubus discolor*), and the evergreen blackberry (*Rubus laciniatus*). The trailing or wild blackberry is the only native species to the Lower Mainland of BC, but is usually outnumbered in the urban habitats by the other introduced species.

All four species share similar characteristics in that they rapidly colonize sites and are pioneer species. They also have very spiny stems. The leaves are toothed and compounded into three to five leaflets. Both young leaves and stems have a woolly or velvety texture of finely matted hairs. All four species produce whitish-purple solitary flowers during May and June and fruit in July and August. The leaves often persist through mild winters.

Blackberry brambles or thickets can be found in a variety of locations. Blackberry species thrive in well-drained soils, which are either neutral or slightly acidic. They are generally shade intolerant, and prefer open or slightly shaded exposures. Evergreen and Himalayan blackberry form dense linear thickets and are common along highways. They can also occur as understorey growth and in openings of deciduous and mixed species forests and at the borders and edges of streams and waterways. One of the most natural habitats for blackberries is at the edges of urban forests. They are usually well represented in the transition zone between the woodlot or linear forest and the adjacent grassland or developed areas.

Blackberry brambles can be cultivated as a hedge. The forbidding prickles and impenetrable thickets form effective natural fences on private or commercial properties.

Since blackberry shrubs can spread their roots quickly and extensively (they have a low root-to-shoot ratio) they are able to utilize soil moisture and soil nutrients more effectively than either trees or grasses. Because of this low root-to-shoot ratio, blackberry draws fewer nutrients and less moisture into the above ground growth and more into the root systems. These characteristics provide a competitive advantage over trees and grasses where soil moisture and nutrient regeneration are irregular. Their perennial nature also allows immobilization of limiting nutrients and, thus, further decreases nutrient cycling, reducing competition and increasing shrub propagation (Smith 1977).

Blackberry brambles are also very efficient soil stabilizers by preventing soil erosion and degradation. Their extensive root systems and prolific growth, even on barren and marginal soils, can prevent further loss of soil especially on moderate to steep gradients. They are very useful for soil conservation in areas undergoing ecological restoration.

Since blackberry is a pioneer species and an effective colonizer of disturbed habitats, it tends to occur in association with other vegetation also proficient at colonization. Typical associations with bracken fern (*Pteridium aquilinium*) and fireweed (*Equilobium agustifolium*) are common, as are associations between blackberry species themselves.

Because most thickets are dense and impenetrable, the deep shade beneath them usually discourages understorey growth. They have a lower energetic and nutrient investment in above ground biomass than trees, therefore, they usually out-compete the latter in the short- to mid-term successional stages (Smith 1977).

European blackberries are invasive and may need to be cut back when re-establishing native vegetation in an area.

These shrubs provide excellent cover and forage potential for a wide variety of animals. The dense, nearly impenetrable thickets provide ideal nesting areas for many birds, such as robins, pheasants, and finches, to name a few. Blackberry brambles are also the favoured habitat of other such animals as coyotes and raccoons. The blackberry communities along forest edges and perimeters of fields may also support their own distinctive animal life. Some species, such as Northern Bobwhite (*Colinus virginianus*), eastern cottontail (*Sylvilagus floridanus*), and Ruffed Grouse (*Bonasa umbellus*) depend heavily on these shrub communities and may disappear if they are destroyed (Smith 1977).

Blackberry shrub communities are very hardy and resistant to many forms of disturbance. Since their dense impenetrable thickets usually discourage significant amounts of human traffic, they are a favoured species for perimeter vegetation surrounding residential or commercial properties and other areas sensitive to human intrusion.

In areas where forest is the normal end of succession, blackberry shrubs, among others, can form stable communities that will persist for many years. Therefore, if successional tree growth is limited or curtailed by herbicide use or by cutting, the shrubs can eventually form a closed community resistant to further invasion by trees. This has wide application in the management of power line rights of way and other urban utility right of way corridors (Egler 1953; Niering and Egler 1955; Niering and Goodwin 1974).

Hedgerows: Linear Thickets

A hedgerow is a closely-planted row of trees, shrubs, and perennials that are in a natural linear arrangement. They are often found along ditches or field margins on farmland (Link 1999). Hedgerows are an important landscape feature in Europe and North America, where they originally functioned as living fences and have developed into corridors that provide ecological, recreational, and cultural benefits (Burel and Baudry 1995).

In Europe, hedgerows have been in place for centuries. Much is known about how they interact with the environment. Hedgerows provide firewood, inhibit erosion, block passage of grazing animals, maintain property boundaries, provide wildlife habitat, and moderate wind and other environmental factors (Forman and Baudry 1984; Fritz and Merriam 1992; Burel and Baudry 1995). For example, 67% of all Irish birds rely on hedgerows for nesting, feeding, and other activities (www.iwt.ie/hedges.html August 16, 2000). In Great Britain, 80% of woodland wildlife, 67% of lowland terrestrial birds, and 67% of lowland mammals use hedgerows for breeding (Forman and Baudry 1984). Hedgerows also have great moderating effects on the landscape. Evaporation is reduced for a distance of 16 times the height of the hedgerow and wind speed is reduced for a distance of 28 times the height of the hedgerow (Forman and Baudry 1984). Soil erosion and temperature are reduced, water quality is improved, and animal biodiversity is increased when compared to fields with no hedgerows (Forman and Baudry 1984; Burel and Baudry 1995). Recently, hedgerows have been thought of in terms of greenways, instead of just as living fences or windbreaks (Burel and Baudry 1995). Using hedgerows as greenways is one way to connect the increasingly fragmented landscape.

This new thought has prompted more research on measures of the quantity and quality of hedgerow habitat. Currently, large studies in England and Ireland are examining hedgerows with this in mind (Halliwell 2000; www.iwt.ie/hedges.html August 16, 2000). Although the studies collectively evaluate hedgerows, methods used are quite variable between studies. In one Irish study currently being completed, the methodology employed in a similar study in the early 1980s is being used. It involves local people surveying their own hedgerows and reporting back to the Irish Wildlife Trust (www.iwt.ie/hedges.html August 16, 2000). The English study (Halliwell 2000) compared aerial photographs taken in 1975 and 1992. Hedgerows present in 1992 were traced onto a grid overlay. The overlay was then superimposed on the 1975 photos and a direct comparison was made with what was not outlined on the grid. The study found a decrease in hedgerow area over time. The amount of the decrease is not reported but apparently varies greatly between study areas. This study recommended that these results be verified by ground surveys, but this was not completed to date. Burel and Baudry (1995) used a combination of field data and aerial photos to characterize landscape units in France. Their aim was to develop guidelines for landscape design and to re-integrate hedgerow network design and the multi-purpose role of greenways. They characterized hedgerows by the type of tree management, that is, how the trees in each hedgerow were pruned. All three of these studies included a survey of the local population of farmers and landowners.

Hedgerows are important wildlife habitat and, while many studies have been done on hedgerows in Europe, only a handful of studies have been done on our local hedgerows. Butler and Campbell undertook a study in 1987 and found there were 29 species of songbirds breeding in the hedgerows regularly (Butler 1999). In the winter of 1990–1991, Butler (1999) studied shorebirds and songbirds in Delta during the non-breeding season. He found that there were 45 species of songbirds using the hedgerows in Delta. Of these, 28 species were found in both tree and shrub hedgerow and all but two species were found in tree hedgerow. Several species of birds of prey were also found using the hedgerows. The birds were probably attracted to the hedgerows for food, perches, and roost sites (Butler 1999).

Hedgerows are planted for songbirds next to greenhouses to compensate for the open field habitat lost when the ground is put under glass.

it seems that in most cases the quantity and the height of trees in a hedgerow positively correlates to the abundance of bird species (Green et al. 1994 and MacDonald and Johnson 1995). In some cases, the relationship does have an upper limit. MacDonald and Johnson (1995) found that this relationship was true for hedgerows up to 4 m tall.

The abundance of woody species and presence of mature trees are also positively correlated to the abundance of bird species in hedgerows (Green et al. 1994 and MacDonald and Johnson 1995). One explanation for this is that not only do they supply food in terms of fruit for birds, they also have a higher abundance of insects. This, in turn, supplies food for another subset of birds, thereby increasing bird diversity (MacDonald and Johnson 1995).

Hedgerows with a high proportion of gaps are negatively correlated with overall abundance of birds and with the number of bird territories found (MacDonald and Johnson 1995). No species of bird was positively affected by gaps in the hedgerow.

If using hedgerows as a habitat compensation strategy is going to be effective in Delta, hedgerows should continue to be planted with high biodiversity and a high abundance of woody plants, an effort to keep the hedgerows with mature trees intact should be made, and gaps should be filled to create a continuous matrix of habitat.

Nature's Services Provided by Shrub Communities

Shrub communities provide many ecosystem services. Their associations with other ecosystem types, such as forests and grasslands, make the services they provide very important. As with forest communities, shrubs help conserve soil. These communities prevent soil erosion (Daily 1997) and are part of the nutrient cycle by recycling nutrient back into the soil. Shrub communities conserve plant, wildlife, and invertebrate biodiversity, as well as genetic diversity. Air is filtered and cleaned by shrubs and other herbaceous plants, and the micro-climate is regulated. Noise is also reduced in the area, as dense shrub communities can act as a buffer.

8

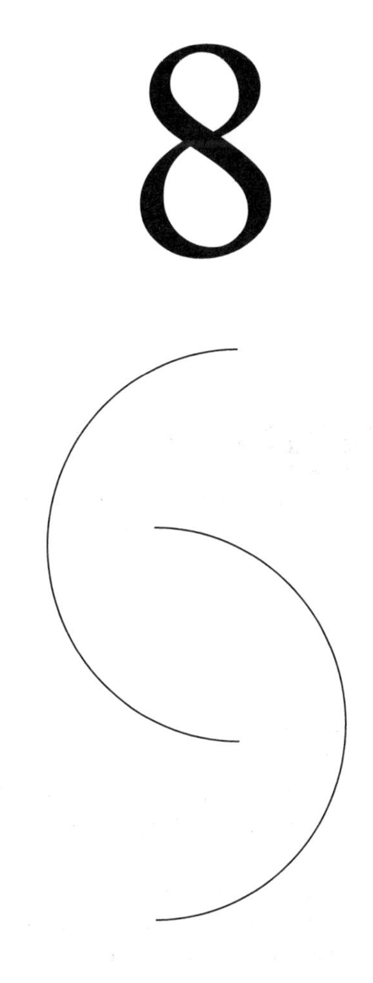

Freshwater Landscapes

Freshwater landscapes range from ponds to estuaries and ditches to rivers. As humans rely on freshwater for many uses, including drinking, irrigation, and hydroelectric power, these landscapes are extremely valuable.

Ponds

Artificial ponds, with an area of less than 6 m^2 (Gilbert 1991), are much smaller than natural ponds. These ponds are often lined with cement or polyethylene and stocked with ornamental plants and animals. Artificial ponds are commonly found in parks, gardens, residential yards, and in front of public buildings.

Natural ponds are larger depressions where water collects. Ponds, like lakes, can be formed through glaciation, volcanic activity, tectonic movement, landslides, and human activity. Old, undisturbed areas are more likely to have a pond than a lake because of lake succession, a process known as eutrophication.

A natural pond has a more diverse community because it is larger and most species present are self-propagating (Gilbert 1991). Ornamental ponds often contain introduced plants and fish. In an ornamental pond only the smaller species, like algae, insects, and some amphibians have enough room to reproduce. Introduced fish, like carp and goldfish, seldom reproduce and their levels are maintained through reintroduction.

The perception of ornamental ponds as less biologically important is incorrect. In an urban environment these "wetlands" provide living space for many native species. Frogs and salamanders can live in ornamental ponds and their concentrations may be higher in these ponds than in the wild (Gilbert 1991). Insects may find these ponds to be the only available habitat for breeding and native plants may be introduced by visiting birds.

The characteristics of water in ornamental ponds vary according to the water source, aeration, concentration of plants and animals, filtration, and maintenance. The water is often supplied from a tap and can have a variable pH. Aeration through waterfalls, fountains and filtration can greatly increase dissolved oxygen levels. The concentration of wildlife will affect nitrogen, nutrient, and organic matter levels; large wildlife populations may foul the water with faeces and dead plants will rot, increasing the level of organic matter. Filtration will reduce nitrates and dead organic matter and may increase oxygen levels and control algal blooms.

Fish, amphibians, and invertebrates are the most common pond animals. In natural ponds trout, catfish, minnows, and feral carp are most common, while in artificial ponds goldfish (*Crasimus auratus*) and many other introduced fish may be present. Frogs and toads are common in both pond types and both provide breeding opportunities for insects such as mosquitoes and dragonflies.

Native aquatic and marsh plants are prevalent in natural ponds and may include common cattail (*Typha latifolia*), small-flowered bullrush (*Sciurpus microcarpus*), water lilies, and water plantain (*Alisma plantago-aquatica*). Ornamental ponds often contain all of these, as well as other non-native species.

Lakes

Lakes are depressions or low points in terrain that collect water and are formed by many geologic processes. Lakes are characterized by three distinct physical regions:

- Littoral zone: a region of shallow water with rooted plants.
- Limnetic zone: open water beyond the littoral zone.
- Profundal zone: lake bottom below the limnetic zone.

A lake's appearance differs from ponds and marshes by the presence of all three regions. Lakes are deeper, allowing for regions of open water and a lakebed with no rooted plants. This differentiates lakes from ponds, which typically have rooted plants throughout. Lakes are drained by streams and rivers (Brewer 1988).

As with ponds, lakes are formed through landslides, volcanic activity, tectonic movement, river activity, glaciation (Wetzel 1983), or by humans. In urban areas, most lakes are found in glacial depressions, or are artificial and may be maintained by dredging. Old gravel pits are often flooded and then marsh plants are introduced. Urban lakes can be found in association with most other urban habitats. In urban parks, lawns often surround lakes, whereas in urban forests trees surround the lake.

Lakes vary with age and location and are characterized by factors such as dissolved oxygen content, depth, stratification, pH, nutrient content, and dead organic material. The two major classifications of lakes based on age and successional status are:

- Eutrophic lakes: These lakes are generally older, shallow, warm lakes with a low dissolved oxygen content and large amount of dead organic material.

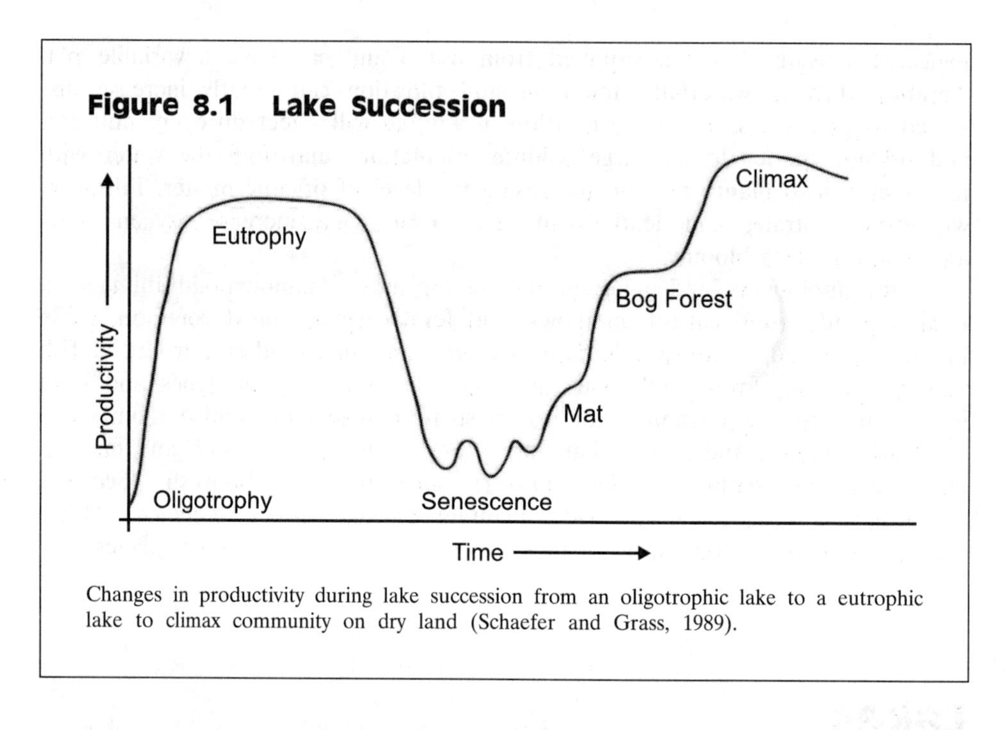

Changes in productivity during lake succession from an oligotrophic lake to a eutrophic lake to climax community on dry land (Schaefer and Grass, 1989).

- Oligotrophic lakes: These lakes are younger and tend to be deeper and stratified, with the lower temperatures usually resulting in a higher dissolved oxygen content. The oligotrophic lake has less organic matter and silt.

Eutrophic lakes, the most common in urban areas, are very productive. The large and diverse plant and plankton communities found in these lakes results in large amounts of dead organic material and much decomposition. Decomposition reduces oxygen levels and restricts fish diversity, but also increases opportunities for insect larvae and waterfowl, such as ducks and coots.

A eutrophic lake is shallow and typically has an extensive growth of lilypads.

Urban eutrophic lakes usually have a dissolved oxygen content below 7 ppm and a somewhat acidic pH. Nitrate and phosphates may be above 0.4 ppm. These conditions often cause visible algal blooms.

Oligotrophic lakes tend to be much larger than eutrophic lakes and are not usually found in urban areas. The impact of varying habitats on an oligotrophic lakeshore is quite small compared to the impact habitat has on a smaller eutrophic lake.

Oligotrophic lakes are less productive, but more diverse. Their higher dissolved oxygen content supports more benthic and/or open water communities (Brewer 1988). The water in oligotrophic lakes is usually clearer than in eutrophic lakes because of fewer and smaller algal blooms. Fish such as trout need higher oxygen content and are more common in oligotrophic lakes. The relative lack of aquatic life also results in fewer waterfowl using these lakes to breed or feed. These lakes may resemble eutrophic lakes at the shoreline if the littoral zone is large. These plants would then eutrophy the oligotrophic lake's edge.

Oligotrophic lakes generally have dissolved oxygen levels between 7 ppm and as high as 12 ppm with a pH neutral to slightly alkaline. Nitrate and phosphates are lower than in eutrophic lakes and there is little, if any, algae.

Urban lakes support a diverse community of plants, phytoplankton, fish, insects and insect larvae, reptiles, birds, and mammals. Plants found in urban lakes include lilies, cattails, iris, and sedges. Lakes are home to many insects, including toe biter water bugs, water striders, and water boatmen. Insect larvae that inhabit lakes include dragonflies, mosquitoes, mayflies, and some dipterans (flies). Introduced fish to urban lakes include rainbow (*Salmo clarki*) and cutthroat trout (*Salmo gairdneri*). Resident fish may include carp (*Cyprinus carpio*), brown bullhead catfish (*Ictalurus nebulosus*) and sticklebacks (*Gasterosteus sp.*). Many native and introduced reptiles are found in urban lakes, including painted turtle (*Chrysemys picta*), red-eared turtle (*Pseudomys scripta*), common garter snakes (*Thamnophis sirtalis*), and red-backed salamanders (*Plethodon cinereus*). Birds that frequent lakes vary from Mallard (*Anas platyrhynchos*) and Canada Goose (*Branta canadensis*) to Barn Swallow and Cliff Swallow (*Hirundo pyrrhonota*) to bathing Crows and American Coot (*Fulica americana*). Raptors, such as hawks and Bald Eagles (*Haliaetus leucocephalus*), use the lake to hunt open water fish like trout. Mammals use urban lakes as well. Raccoons may hunt along the shoreline, while muskrats (*Ondatra zibethica*) and beavers (*Castor canadensis*) take up residence at stream inlets and outlets.

The yellow pond-lily (*Nuphar polysepalum*), native to coastal British Columbia, is a common lake species found in shallow water (Klinka et al. 1989). The non-native white fragrant waterlily (*Nymphaea odorata*) is also common to shallow areas. Plants common around the shores of lakes include the Sitka sedge (*Carex sitchensis*), which is an indicator of nitrogen rich organic soils, and the common cattail, which can be seen in many lakes and is an early indicator of eutrophication (Klinka et al. 1989). Water plantain is a common shallow rooting plant found in muddy ground. Often found with the water plantain are cattails, bulrushes, and arrowheads (*Sagittarai latifolia*) (Fish Habitat and Enhancement 1990).

Marshes

Freshwater marshes are typically wet most of the year. Even during the dry summer months, there is enough moisture to keep the ground soft and muddy and vegetation growing. Skunk cabbage (*Lysichiton americanum*) dominate shady coastal marshes. Cattails, sedges, reeds, and bulrushes generally dominate the marshes that are exposed to full sun.

Freshwater marshes are generally found adjacent to slow moving streams or creeks or form the shallower portions of ponds and lakes. They can also occur at drainage basins of underground springs. Some marshes occur in depressions along highways, often functioning as storm drainage basins.

Freshwater marsh vegetation is typically grass-like, upright and tall. Plants like cattails, rushes, reeds and sedges have long and narrow leaves with few or no branches. They have hollow and fibrous stems that enable them to withstand long periods of being submersed under water. They possess extensive networks of rhizomes with which they propagate vegetatively. Rhizomes penetrate the muddy substrate and are capable of extending towards the water's edge. Once colonization of the water's edge begins, the flow of water decreases, and sediment accumulates. Along with this process, the accumulation of organic debris on the mud due to the slow decomposition of the stems and leaves further increases the level of the marsh and extending the drier upland habitat.

Skunk cabbage is found in damp, nitrogen rich soils assoicated with marshes and seeps of ravines

Freshwater marshes are important habitats for various mammals, birds, amphibians, and reptiles that utilize these areas as shelter and a food source. Red-winged Blackbird (*Agelaius phoeniceus*) and Yellow-headed Blackbird (*Xanthocephalus xanthocephalus*) can be found displaying their courtship behaviours and also feeding on the cattail flowers. Great Blue Herons often sit and wait for prey such as frogs and snakes. Waterfowl such as Mallards, Gadwall (*Anas strepera*) and American Widgeon (*A. americana*) are voracious grazers of roots and rhizomes. American Coots build their nests amongst the marsh vegetation. Marshes are also frequented by raccoons (*Procyon lotor*), muskrats (*Ondatra zibethica*), weasels (*Mustela sp.*), opossum (*Didelphis marsupialis*), and shrews during their foraging activities.

Flowering plants like the water plantain, skunk cabbage, purple loosestrife (*Lysimachia salicaria*), arrowhead, and American brooklime (*Veronica beccabunga*) often grow with the cattail-sedge-rush communities. Hardhack, salmonberry thickets, willow, and red alder are often found growing on the drier portions of the marshes where exposure to water is minimal.

Bogs

Bogs are mineral and nutrient-deficient, acidic environments that support interesting plant communities adapted to survive in such extremes. They are dominated by *Sphagnum* moss. Bogs are usually in depressions, but some domed bogs have raised centres and are surrounded by a depressed margin, called a lagg (Agriculture Canada 1976). Domed bogs are of the low shrub type, with dense growth of Labrador tea (*Ledum groenlandicum*), bog laurel (*Kalmia microphylla*), cranberry (*Oxycoccus oxycoccus*), and salal, along with stunted shore pines (*Pinus contorta*).

Bogs are formed through the terrestrialization process of a lake, known as hydric succession. The process begins with the infilling of a lake by sediments and colonization of the wetland vegetation. Over time, organic matter accumulates and gradually blocks the flow of incoming water. The area then becomes isolated from a nutrient supply. With the increase of plant growth, the organic substrate of available nutrients is eventually depleted, resulting in a decrease in productivity. Further accumulation and decomposition of the detritus matter causes the depletion of the available oxygen. The incoming water is derived solely from rainfall. Bogs are dependent on rain and airborne detritus particles for nutrients. Rainwater, however, is nutrient poor. This environment then becomes oligotrophic (deficient in plant nutrients) and anaerobic (lacking oxygen), which is a suitable habitat for *Sphagnum* moss. It can tolerate the marginal nutrient regime and its own physiology maintains low pH (Moore and Bellamy 1974). *Sphagnum* also maintains the water table at the growing surface by capillary action.

Any surface water feature such as a pond, lake or marsh is especially valuable in supporting urban biodversity.

Sphagnum species dominate bog environments, which help in retaining a unique microclimate (Taylor 1990) that can act as refuge habitats for the typical bog plants that are not normally found at this latitude and climate regime (Kringa 1969). These plants are able to survive without any interference and competition from other terrestrial vegetation species.

Bog plants are successful because of their adaptation to nutrient- and oxygen-poor environments. Shrubs such as bog-rosemary (*Andromeda polifolia*), bog-laurel, and labrador tea cope with the environmental extremes by developing characteristic growth habits and forms. They have thick, shiny evergreen leaves with heavy cuticles, recurved margins, woolly under surface, and/or sunken stomata for protection against the loss of water through evapotranspiration.

Plants like the round-leaved sundew (*Drosera rotundifolia*) are insectivorous and rely on animal proteins as a dietary supplement. They have leaf blades that are covered with hairs that secrete a sticky substance to trap insects.

Many species of bog plants, such as cloudberry (*Rubus chamaemorus*), are rare. These are tundra plants that are found growing in the southernmost periphery of their range.

Bogs have undergone unique disturbance regimes. Peat mining and drainage for development and agriculture are two main disturbances that are common in bogs and detrimental to these ecosystems.

Bogs are unique natural phenomenon. They are capable of harbouring unique plant species that are normally found in tundra and boreal climates. The bogs developed over the centuries to become refugia for these specialized plants.

Streams and Rivers

Streams and rivers function as water discharge areas for watersheds. Water is collected in the watersheds and is filtered through the permeable soils and rocks into the groundwater. Here, the groundwater flows above the impermeable layers of rocks or clay until becoming exposed as the headwaters of a stream channel. As the water flows down the topography of the watershed, it cuts through various morphological features, forming ravines and valleys.

Generally, streams and rivers are fed by rainfall in the winter and snow and glacial melt throughout the summer. They generally have a continuous water supply. It is different for streams and rivers in urban areas, where the majority of the headwaters may be paved and storm drains may be a dominant source of water. Following rainfall or a storm, water flow is usually greater; sometimes overflowing the stream banks and damaging the riparian habitats through "flash" floods. In dry summer months, water levels may fall considerably due to the decrease in groundwater recharge.

Streams and rivers can be found flowing down hillsides, mountainous slopes, and along ravine and valley bottoms. In urban areas, many streams may have been culverted and paved over. Some urban streams in ravines that have very steep banks are left in their natural state because they are too hazardous for development. Such areas are an important habitat for wildlife in cities.

A stream or river is unique to the watershed it drains. The physical and chemical characteristics of a stream are determined by the characteristics of the drainage area. Rock and gravel predominate in streams with a steep gradient, whereas streams with relatively low gradient tend to accumulate silt and sand. From headwaters to the mouth, streams flow from higher to lower gradient. Along with the change of gradient, the composition of the streambed becomes finer, the channels wider, and the stream banks lower.

There are two types of habitats associated with streams: riparian along the shore, and the streambed. Both are important components in the ecology of streams.

Stream systems are dependent on the vegetation that grows along and above the stream. Riparian vegetation functions in filtering pollutants, stabilizing the bank, and providing shade and important nutrients. The leaves of the streamside vegetation fall into the stream and provide food for aquatic insects and release proteins during decomposition. In the summer, terrestrial insects that live in the streamside vegetation fall into the streams and become an important source of food for fish. Riparian zones are also important habitats to other wildlife such as shrews, voles, mink (*Mustela vison*), muskrats, weasels, Great Blue Herons, Belted Kingfisher (*Ceryle alcyon*) and numerous songbirds that utilize the canopies of the trees and shrubs for feeding and their nesting.

The substrate of streambeds is important in determining the productivity of a stream. The different types include bedrock, boulder (>25 cm diameter), cobble (5–25 cm), gravel (0.2–5 cm), and fines (<0.2 cm). Each indicates characteristic velocity of the stream's current. Also important is the degree of embeddedness of the substrate material in the streambed. Gravel and cobble that is well embedded in the streambed may not be as useful habitat compared to situations where the gravel and cobble are less embedded. Each type of substrate is capable of supporting its own unique biotic community. In streams with bedrock and boulder as substrate material, stream velocity is extremely high. This environment has a very limited nutrient supply, and therefore contains little aquatic life. Mayfly nymphs and nematodes are found here.

One method to stabilize a stream bank is to create a retaining wall or wattle of woven branches with species that will sprout from the branch nodes, such as willow and red-osier dogwood. Wattling is a form of bioengineering and results in a living wall that creates wildlife habitat as well as preventing erosion.

Other stream characteristics relevant to supporting biodiversity include the volume of water discharge, instream cover (pieces of large woody debris, rooted cutbanks), off channel habitat, overhead canopy, bank stability and temperature (Table 8.1).

There is much debate about the required width of buffer zones around streams to protect their biodiversity from nearby urban development. A buffer zone of 30 m from top of bank is recommended to protect the productivity of

Table 8.1 Relationship of Representative Stream Life to Water Temperature

Temperature Range	Types of Stream Life
20–25° C (warm)	Plant life high; fish disease risk high; warm water fish (bass, carp, crappie, catfish, bluegill); caddisflies; dragonflies
13–20° C (cool)	Some plant life; moderate fish disease risk; oxygen loving fish (trout, salmon, sculpins); stoneflies; mayflies; caddisflies; water beetles; water striders
5–13° C (cold)	Some plant life; low fish disease risk; oxygen loving fish (trout, salmon, sculpins); stoneflies; mayflies; caddisflies

(**Source:** Stewardship Series 1996)

salmon bearing streams (Stewardship Series 1994). This extent of a vegetated riparian zone would help to shade the water, keeping temperatures lower, and would provide cover and food for juveniles. However, for salamanders it may well be that a buffer zone around a stream would not help to maintain their populations, no matter how wide the zone may be (Wilson and Dorcas 2003). Instead, conserving salamander populations in headwater streams requires land use planning throughout an entire watershed. Establishing relatively narrow riparian buffer zones would be ineffectual in this case.

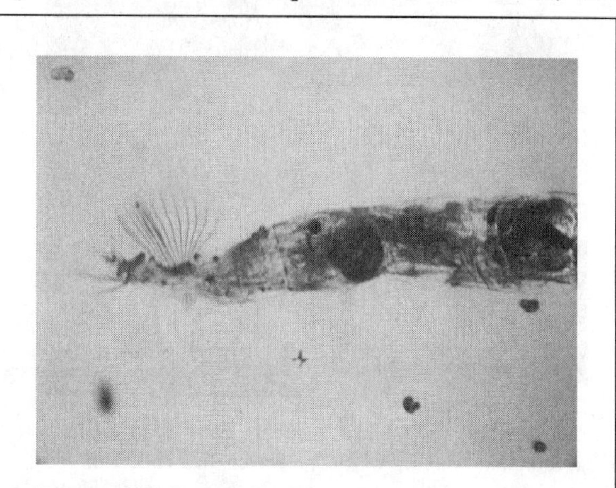

The caddisfly larva, about 3 cm long, is an indicator of good water quality.

The boulders and cobble in urban streams are larger at the headwaters and smaller closer to the mouth; progressively heavier water flows downstream wash out larger rocks.

Cobble and gravel stream beds also have high stream velocity and are highly oxygenated and capable of carrying food supply. Algae such as blue-green algae and *Fontinalis* moss grow on the gravel and are the primary food source for aquatic invertebrates. These invertebrates have adapted to the life in a fast flowing stream. Such adaptations include having flat bodies allowing them to cling to rocks and possessing powerful suction discs on their undersides. Mayfly, stonefly, and caddisfly larvae are found in streams with such substrate. They are voracious predators capable of capturing food particles, such as diatoms, in swift current. These larvae are an important food source for juvenile salmonids. Some of the fish found here are cutthroat trout, minnows, stickleback, and juvenile Coho salmon (*Oncorhynchus kisutch*). Streambeds with sandy substrate have low productivity. Most algae and moss are incapable of growing without adhering to rock surfaces. Food is often washed downstream. Some animals that may inhabit this type of environment are nematodes, caddisfly, and mayfly larvae.

Silty and muddy stream beds are found in areas with low velocity, which causes sedimentation to occur. Productivity in these areas is high. The diversity of primary producers is high, which includes water weeds, arrowheads, water plantain, and cattail. The diversity of aquatic invertebrates is also increased.

Large woody debris in streams plays an important role in maintaining stream function for salmonids. A log in the stream that restrains the flow of water can create a gravel bench of the upstream side of the log, good spawning habitat for salmon, and a plunge pool on the downstream side, a good rearing site for juveniles.

A healthy stream will have natural meanders, which naturally occur a distance equal to twelve times the width of the stream for the full "S" meander that typically contains a pool alternating with a riffle, followed by another pool and riffle. Associated with this would be quiet pools and turbulent riffles (small rapids), that support a high degree of biodiversity.

Many times urban streams are cleaned out of woody debris in an attempt to beautify the view, and the stream is channelized or straightened. These practices reduce the number of species a stream can support.

Algae and detritus are the primary producers in the food chain. Aquatic invertebrates such as insect larvae, aquatic insects, leeches, snails, and worms are primary consumers. They, in turn, become food for fish, amphibians, reptiles, and birds, which are predated on by larger animals.

Estuaries

An estuary, as defined by Pritchard (1967), is a semi-enclosed coastal body of water which has a free connection with the open sea, and within which sea water is diluted with fresh water derived from land drainage. It is generally located at the lower end of a river delta. Brackish marshes may be found along the bank of the river delta where the river water is affected by the salinity of the tidal sea water.

Estuaries are important components of a watershed system. They essentially function as nutrient traps and occur at a land-water interface, which is greatly influenced by salinity and inundation of water during tidal exchange. Nutrients are brought in from the marine environment, the terrestrial environment, and the watershed environment. The trapped nutrients are distributed throughout the estuary by the natural phenomena of tides, wind, river current and the salt wedge. The salt wedge is created by the seeping action of salt water along the river bottom, and results in an upwelling of nutrients. Because of these nutrients, estuaries support a diversity of wildlife from waterfowl, shore- and water birds, raptors, mammals, and fish.

The soils of the estuary are deltaic in origin. Rivers carry alluvial deposits through watersheds and deposit them at the estuary.

There are three types of habitat associated with an estuary: benthic, intertidal, and upland. Marine plants such as seaweeds may grow on stable mud flat surfaces common in the benthic zone. Common eelgrass (*Zostera marina*) is commonly found growing in sand permanently submerged in water. Eelgrass occurs in the lower intertidal zone of many estuarine habitats with salinities of 10 to 30 parts per thousand and temperatures of 10–20°C (Fish Habitat Enhancement 1990). It is important in sediment stabilization and provides excellent feeding habitat and cover from potential predators. Pacific herring (*Clupea harengus*) migrate into the estuaries during the reproductive period of their life cycle. The faunal community of this habitat include snails, clams, sea stars, isopods, amphipods, copepods, and crabs (Fish Habitat Enhancement 1990). Gunnels (*Pholis sp.*), sculpins (*Icelinus spp.*), rockfish (*Sebastes sp.*), cabezon (*Scorpaenichthys marmoratus*), bay pipefish (*Syngnathus griseolineatus*), greenlings (*Hexagrammus spp.*), tubesnout (*Aulorynchus flavidus*), and sole are among the common residents here that take advantage of the eelgrass meadows for food and cover. The intertidal zone is an important habitat for the juvenile salmonids. They spend some time here to feed and acclimatize their physiology prior to their oceanic migration. Canada Goose, American Widgeon, Northern Pintail (*Anas acuta*), Canvasback (*Aythya valisineria*), Mallard, and Green-winged Teal (*Anas crecca*) are important herbivores of the eelgrass. Northcote (1952) found the mysid *Neomysis* to be abundant in the arms of the lower Fraser River (estuary portion). Hallam (1973) found hydrozoa, errant, two to three species of polychaetes, oligochaetes, mussels, clams, shipworms, snails, barnacles, amphipods, a decapod, dipteran larvae and nematodes. Some of the halophytic (salt tolerant plants) invertebrates, such as shipworms (*Bankia setacea*), are capable of survival in a brackish (mixture of salt and freshwater) environment. They inhabit depressions near pilings, which do not come in contact with freshwater during tidal exchange.

Eelgrass (Zostera sp.) is an important pant in marine estuaries and is planted to restore disturbed habitat.

The intertidal zone is where land is exposed to open air at low tide and submerged at high tide. Estuarine marshes are predominately colonized by Scirpus and Carex. The distribution of these plants is dependent on length of tidal flooding, drainage and, most important, water salinity. Moody (1978) found that the distribution of Scirpus and Carex is dependent on the relationship of the elevation-tidal level. Cattails are found in areas with high freshwater influence. Various studies (Bradfield and Porter 1982, Hutchinson 1982) on the estuarine marshes have reported that the marsh community can be grouped into distinct vegetation community zones: low, middle and high marsh zones. These three zones have varied exposure to the tidal flooding.

Upland zones are typically wooded. The dominant trees in coastal British Columbia are bigleaf maple, red alder, mountain alder (*Alnus tenuifolia*), Douglas-fir (*Pseudotsuga menziesii*), and western redcedar (*Thuja plicata*) (North et al. 1984). These areas are often dyked and converted to agricultural land. Some of the grassland communities found in areas like the Sea Island are typical of the primary and secondary successions (Hoos and Packman 1974). Some of the more abundant grasses are reed canary grass (*Phalaris arundinacea*), perennial ryegrass (*Lolium perenne*), creeping bentgrass (*Agrostis sp.*), and *Fescue*. Waterfowl such as Canada Goose, Snow Goose (*Chen caerulescens*), Mallard, Northern Pintail, American Wigeon, Green-winged Teal, Cinnamon Teal (*Anas cyanoptera*), and Gadwall frequent these areas to feed on the grass rhizomes, especially during migrational stop-overs.

Ditches

There is a wide range of drainage ditch habitats in the urban landscape. Ditches vary in biological importance depending on their location and size. These factors also determine the differences in their aesthetic appeal. Ditches may be small and narrow furrows along roadsides, or wide and steep canals near industrial sites. The majority of ditches are small and cannot support as much biodiversity as more complex ecosystems such as canals and marshes, but they do contain grasses and insects that are important sources of food and shelter for amphibians and some birds. In small drainage waterways, scenic value is greatly reduced because they contain less vegetation, and are limited in size and species diversity.

Larger drainage ditches do not have a gradual or well-defined transition zone between the water's edge and the grass bank borders. Often, as is the case with most canals, the change is very abrupt. The canals and marsh drainage ditches enhance the natural environment by attracting wildlife diversity. For this reason these important waterways offer potential scenic value.

Drainage ditches and their associated grass bank habitats are less likely to occur in larger metropolitan areas because they are usually replaced by a more sophisticated drainage system or the water flow is contained in and directed by a network of culverts. Drainage ditches and small run-off streams are thus more likely to be found in less developed municipalities and suburban centres. Ditches in urban areas often occur on flood plains. Drainage ditches in suburban areas are sometimes left after development to prevent flooding during heavy rainfall. These types of ditches are used as floodplain control systems in farm areas. These ditches contain high nutrient levels from the farms' runoff and may resemble small marshes.

Soils in some drainage ditches may contain high nutrient levels. Moist soils with lush vegetation contribute high quantities of organic materials. Because both leaves and roots of grasses die back annually, organic matter builds up and creates a rich humus and essential plant nutrients.

Due to a continual water flow in drainage ditches, sand and silt are carried by the water and settle in different areas of calmer flow. This causes a build up of a mixture containing sand, silt, and clay particles called loam. This mixture of loam and humus has an effective characteristic: it allows rainfall to penetrate into the soil, maintaining moisture for days.

Smaller waterways do not exhibit as much species diversity as canals or marshes because of the smaller habitat areas. Some typical wildlife from these areas may include Great Blue Herons, Mallards, muskrats, frogs and raccoons. The plant species associated with these areas are also almost always limited to grass species and smaller shrubs. Horsetails and ferns are also typical examples of the water loving vegetation in these areas.

Large waterways contain greater species diversity. Some canals and marshes provide salmon rearing habitat and others may provide secluded nesting sites for various species of birds. Some typical examples may include swans, Mallards, Great Blue Herons and many more. These lush waterways also support an array of invertebrate species. Some examples of these may include pond snails, grasshoppers, moths and diving beetles.

A beaver dam. Beavers often thrive in large urban lakes and can become a serious pest.

Exotic plants may escape residential lawns or gardens and can compete with native vegetation for habitat. For example, if purple loosestrife is left to grow unchecked it may cause serious damage, causing an ecosystem to become unstable. Exotic plants often don't have native competitors or predators, which results in a crowding out of native plants and loss wildlife diversity that rely on them.

When drainage ditches are replaced by storm drain systems the natural watershed cycle is distorted. The installation of new drainage systems, such as storm sewers and piped water diversions, causes flooding and erosion in the local ravines that they drain into. Urban areas are usually heavily paved and do not have proper filtering access during rainfall. Chemical pollutants used in residential areas, such as pesticides or other hazardous materials, may enter directly into a fragile watershed from street runoff, inflicting a potential threat of contamination to the plants and animals and destabilize the ravine ecosystem. Less water drains into ravines from drainage ditches. The vegetation on the slopes stabilizes the water content and minimizes erosion. It reduces sediment input to streams and maintains the quality and temperature of the water.

Nature's Services Provided by Freshwater Ecosystems

Freshwater ecosystems provide many ecosystem services ranging from supply of water to supply of other good to non-utilitarian values. For example, freshwater is used for drinking, cooking, washing, and other household uses (Daily 1997). North Americans use an average of 300 L of freshwater for household uses everyday. Freshwater is used in manufacturing and production of goods and other industrial uses including thermoelectric power generation (Daily 1997). Irrigation of agricultural land, golf courses, and parks is another major service provided by freshwater ecosystems.

In addition to supply water for a variety of uses as previously mentioned, freshwater ecosystems also provide a supply of goods including fish and waterfowl (Daily 1997).

- Of the total volume of water on the planet, only 2.5% is fresh, of which 2/3 is locked in glaciers and ice caps;
  - Precipitation on land is estimated at $110,000 km^3/an
  - Worldwide, dams, reservoirs, canals, wells and other forms of infrastructure remove 4,430 km^3/an from freshwater ecosystems
  - Worldwide desalination accounts for 0.1% of water use at a cost of $1–2 per m^3;
  - In all, peoples' freshwater use produced by desalination would cost $3 trillion/an — 12% of 1997 gross world product. If one were to include water provided by evapotranspiration as precipitation on fields and forests, this value would be nine times greater — $27 trillion per year — and would exceed the entire 1997 gross world product;
- The annual freshwater fishery harvest in 1990 was about 14 million tons and was worth $8.2 billion

- The value of sport fishery is at least $16 billion in direct expenditures including equipment, travel, etc. (Felder and Nickum 1992) with total economic output approximately $46 billion;
- Waterfowl hunting in 1991 worth $670 million of direct economic impacts;
- Pollution dilution services in the US is worth at least $150 billion;
- Revenue derived from goods transported on freshwater in the US is worth $360 billion (US Dept of Transportation 1993, 1994) and in western Europe is worth $169 billion (UNEP 1992, UN 1994).

Freshwater landscapes also provide many nonextractive ecosystem services. These nonextractive services are easily taken for granted and are difficult to value since there are no goods bought or sold associated with them. These benefits include natural flood control, pollution dilution and water quality protection, transportation, recreation, and bird and wildlife habitat (Daily 1997).

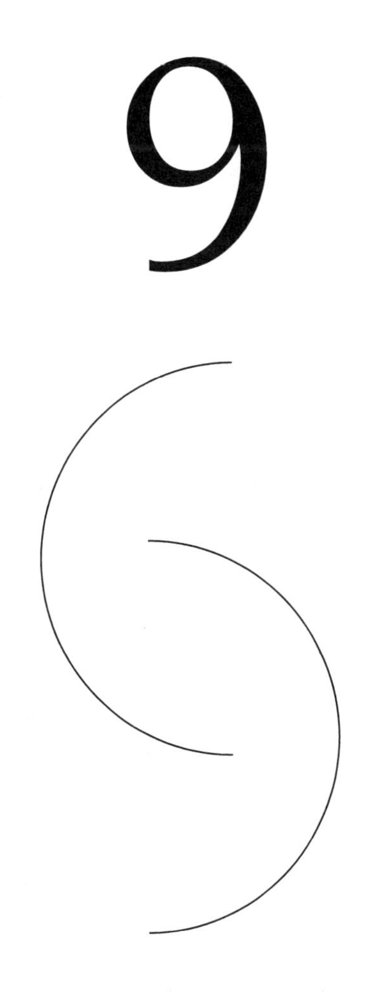

Open Spaces

Open spaces in urban areas include grasslands, agricultural land and old field habitat. Some of the areas are often intensely managed for food production or livestock, while others are left to undergo ecological succession.

Grasslands

A grassland is an expanse of grass, possibly with some scattered trees. Grasslands do not have extensive shrub communities except in small, localized clumps. The grass itself is made up of many species and is of varying height due to its age and species.

Abandoned lots are gradually replaced by tall grasses. These are first invaded by sun-tolerant deciduous pioneer trees, such as birch on more acidic sites and alder on alkaline sites. The site is later dominated by shade-tolerant trees, either deciduous or coniferous, depending on the area.

Urban grasslands are usually found in areas of recently disturbed areas. Because grasslands are one of the earliest successional stages, they grow quickly on land that has been disturbed. Grasses are the first species to colonize areas that have recently been disturbed.

A rare grassland type is the climax grassland: an area of grassland that will remain grassland with no further successional change. Grasslands occur as climax communities in arid regions such as the Central Interior of British Columbia, the prairies in Canada and the Great Plains in the United States.

The range of grassland type and species composition in any region is vast. The nature of grasslands is dependent on the species present. Older grasslands will have a different species composition from young, early successional ecosystems. The identification of grass species can be difficult. In urban habitats the species diversity is lower and less dependent on grass type. This classification is based on the types of flora and fauna found in relation to the grasses. These classes are “grass only”, “grass with shrubs”, and “grass with shrubs and trees”.

Grass-only grasslands are found within a few years of a site being disturbed. When only grass is present, the site is usually in an early successional stage, as only pioneering grass species are present. Wildlife species diversity is low because the nesting opportunities are restricted to animals that nest on the ground, such as the savannah sparrow and ring-necked pheasant.

The addition of shrubs enhances species diversity by bringing another dimension to the site. There are more nesting sites for wildlife and food sources are more abundant. Shrubs like hardhack, salmonberry, and Nootka rose (*Rosa nutkana*) attract insects while providing opportunities for other wildlife.

As trees begin to grow with grasses, the site is able to support a greater diversity of wildlife. Tree nesters such as crows, warblers and robins may find suitable sites here. Trees also add perching sites for raptors like owls that need the open grassland to find voles and mice. Colonization by trees usually signals an end to the grassland as the area has moved through succession towards an urban forest.

Soil is a major determinant of the type of grassland and the speed at which succession will occur. Grassland soils range from wet and acidic to dry and alkaline. The species of grass present is determined by its ability to tolerate the specific conditions. Indicator species may identify the soil type and the successional path of a site. An example is California brome (*Bromus carinatus*), a grass that prefers dry, nitrogen-medium soils (Klinka et al. 1989). On this site it is possible to predict that succession will not be to water preferring deciduous trees like willow.

Agricultural Land

Agricultural fields are open spaces that have been tilled to grow crops. The land is typically uniform, with one crop grown over large areas. Land design and colour are dependent on the crop and its cultivation requirements: corn is planted in parallel rows, cranberries are flooded in trenches, and hops are hung from supports in a characteristic Y-shape.

Farmland is located in areas of high cultivation potential. This means that suitable soil and climate requirements are present. Generally, agricultural fields are found in valleys and floodplains where soil and drainage make for acceptable growing conditions.

Agricultural fields may be divided into two general types based on the steepness of the land: zero or minimum grade, and medium grade.

Zero or minimum grade is where there is a low gradient and soil erosion is minimal. Soil and water qualities are maintained without extreme intervention. Zero or minimum grade land provides an ideal habitat for a large diversity of organisms. Although this land is continually managed, the top layers of soil and organic matter are retained, as is the biodiversity associated with the top layers of soil.

Medium grade is where there is a slope and it is subject to soil damage by water erosion. Land with a well-defined slope is subject to erosion, so species diversity may fluctuate depending on the amount of rich soil available. The erosive action of both water and wind result in organisms found in the top layers of the soil profile being washed down the slope. Many of these micro-organisms are important in retaining high soil quality in terms of nutrient turnover, aeration, nitrogen-fixation, and plant hormone production. Without them, maintenance and functioning of the soil deteriorate.

Soil fertility, a soil's ability to provide essential chemical nutrients for plant growth, is perhaps the most important component of agricultural land. Fertility is affected by a soil's structure, texture and living components. In terms of structure, the topsoil is the most important layer for farm crops because it contains life-sustaining nutrients in their most readily available form. Maintenance of this layer is achieved by mechanical ploughing and fertilizing but deep-rooted crops also contribute by bringing up materials from the subsoil, losing their leaves, or dying, and so help to restore much of the nutrients to the topsoil.

Agricultural fields can be important habitat for threatened or endangered species such as these Sanhill Cranes (*Grus canadensis*).

Soil texture, determined by particle size, plays a fundamental role in water-holding capacity and aeration. Soil water is dependent on the amount and type of clay present, and on the amount of organic matter within the soil. Clay particles, having a diameter of approximately 2 microns, have a large amount of surface area relative to their weight and, therefore, have the capacity to absorb water, other molecules and ions on its surface (Brewer 1989). Organic materials also have this property and cause the soil to aggregate. A soil deficient in clay or organic material has little structure and does not hold together well. If the clay component is too high, the soil becomes hard packed and impenetrable to plant roots.

Soil biota are often overlooked in their importance for maintaining high soil quality. Organisms range in size from burrowing vertebrates to insects to microscopic fungi and bacteria. They are vital in the breakdown and recycling of the chemical constituents without which the soil would stagnate and crop yield and production would deteriorate.

Important organisms such as the common earthworm survive best in moist, loosely packed soil containing a large amount of organic matter (Andrews 1973). Worms are important because they promote soil aeration by creating tunnels. This permits oxygen to circulate throughout the soil, supplying it to the roots and shafts of underground vegetation. Another important role of earthworms is that they pass soil that is rich in minerals up to higher levels, within reach of the surface roots.

Other important organisms that are confined to the top portion of the soil profile are the enchytraeids, or white worms. Their function is to coat moist plant remains in the upper humus with a secreted alkaline solution. When the plant tissues have dissolved, they can easily be eaten and the waste further mixed into the humus (Andrews 1973).

In addition to growing food for people with principal crops such as wheat, market-garden vegetables and small fruits, much cultivated land is planted with feed grains as pasture for livestock.

As mentioned previously, the importance of soil biota is underestimated. Invisible to the naked eye, micro-organisms may, at times, be far more active in producing new growth than we are aware of. The weight of bacteria alone in some soils has been estimated at one tonneor more per acre! (Desman 1976). Soil fungi may be more abundant than bacteria, especially in acidic soil and in

Large greenhouse operations are becoming increasing common on agricultural lands around cities and can destroy old field habitat.

the deeper regions of soil where oxygen levels are lower and carbon dioxide levels higher. Like bacteria, fungi are essential in the decomposition of organic matter such as cellulose and starch and also break protein down, releasing essential nitrogen compounds into the soil (Desman 1976).

Macrofauna of agricultural land are diverse. Although it is true that the majority of organisms are the annelids, nematodes, arthropods, and molluscs, there does exist a population of vertebrates. Other animals that are attracted to agricultural fields include birds, mice and deer, which feed on soil organisms, crop fruits, roots, seeds and leaves.

Old Field Habitat

An old field habitat results from abandoned agricultural fields becoming overgrown with tall grasses. There may or may not be shrubs, hedgerows and woodlots present as well. Old field habitats are used by a wide range of songbirds. They are is important as loafing and feeding grounds for migrating waterfowl and shorebirds.

An old field habitat has an abundance of rodents, especially mice and voles. These are an important food source for many raptors, such as the Northern Harrier and Barn Owl. They are also thought to be an important source of food for young Great Blue Heron.

Nature's Services Provided by Open Spaces

Open spaces in urban areas are often thought of as wasted space, yet they provide many important services. Grasses and flowering plants on grasslands and agricultural land are essential for pollination services. Pollinators comprise all bees, butterflies, moths, flies, other insects, birds, bats, and other mammals, as well as wind and water. They are responsible for pollinating the majority of food and non-food plants throughout the world. Not only do open spaces provide plants to pollinate, they also provide other important habitat components for these pollinators. The following details the importance of pollinators as outlined by Daly (1997):

- society has 1,509 cultivated plant species;
- 1,330 have known pollinators and less than 2% are wind pollinated;
- anywhere from 73–83% are pollinated by bees;
- 19% are pollinated by flies, 6.5% by bats, 5% by wasps, 5% by beetles
- 103–108 crop species provide 90% of people's food globally (of which 71 are bee pollinated);
- in the US in 1985 it was estimated that honeybees improved the value of yields and productivity by $1.6–5.7 billion

- at that time pollination service charges paid to beekeepers by farmers was $81 million.

Open spaces are part of soil conservation, especially if located on a low to minimum grade site. The following observations pertain to the importance of soil in maintaining biodiversity and providing other natural services (Daily 1997):

- under one square yard of pasture in Denmark:
  - 50,000 small earthworms and relatives
  - 50,000 insects and mites
  - 12 million roundworms
- a single gram of soil (pinch) has yielded:
  - 30,000 protozoa
  - 50,000 algae
  - 400,000 fungi
  - billions of bacteria

As with forests, grasses and other species associated with open spaces are able to sequester a high amount of carbon from the atmosphere (Daily 1997) and absorb other harmful pollutants. These ecosystems, especially if not intensely managed, have high genetic diversity. A high genetic diversity allow organisms to be more resilient to stress and disturbances. The following highlights the importance of grasses to people and the biosphere (Daly 1997):

- Grasslands cover about 25% of land surface of the Earth
- Carbon losses to cultivation for crops are between 0.8–2 kg/m^2
- Carbon sequestration is 10,000 kg/ha/an or $200/ha
- Methane uptake is 0.5 kg/ha/an or $0.05/ha
- Nitrous oxide emission uptake is 0.19 kg/ha/an or $0.60/an
- Ameliorate climate through albedo and reduce erosion.

10

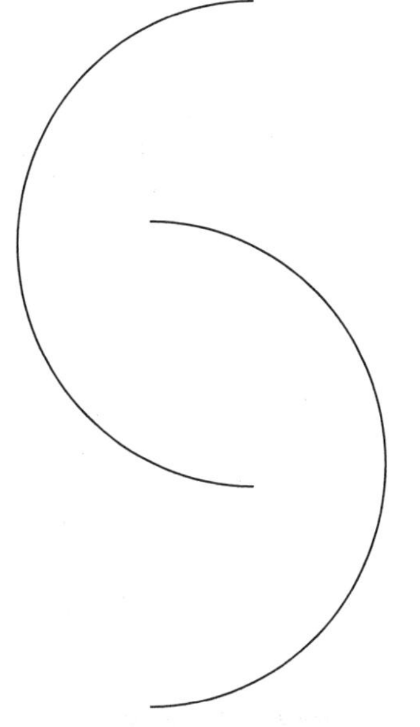

Barren Landscapes

Barren landscapes include everything from bridges to buildings and landfills and vacant lots. They have little vegetation and on the surface, provide very little wildlife habitat.

Bridges

Bridges provide continuous people passage over a body of water, road, or valley. They are constructed from a variety of materials of which wood, steel and reinforced concrete are most common.

Bridges range in length from a few feet or metres to several miles or kilometres and may be classified as single- or multi-span. They are further classified as causeways, overpasses, and viaducts depending on how tall and long the bridge is and whether it passes over water, land, railways, or roads.

Bridge undersides are used as roosting and nesting sites for Barn Owls, bats and Cliff Swallows.

When a bridge crosses over land, the ground under the bridge can either have high or low maintenance. Some sites are regularly maintained to give a clean, manicured appearance. These areas often have public access and offer recreational opportunities, such as bike paths. Areas of low or no maintenance are generally found beneath bridges that have no public access or are difficult to maintain due to physical structure or rough terrain.

The wildlife habitat provided beneath bridges is harsh and often stressed. They are extreme compared with most of the biosphere (Brewer 1988). Conditions beneath the bridge deck are generally drier, cooler, and darker. Soil is typically compacted and consists of sand, gravel, and construction rubble. Such nutrient-poor soils and adverse microclimatic conditions limit the diversity of species. These habitats often see high populations of few species because the habitat range is so narrow (Brewer 1988).

The height and width of a bridge are two important factors determining the presence and abundance of vegetation and wildlife. In general, low bridges (less than 7 m high) display little diversity because of the extreme shading. Bridges higher than 40 m allow more light and greater plant diversity, but height also restricts nesting potential, keeping diversity of wildlife low. Typically, only cliff-dwelling birds, such as the Cliff Swallow, inhabit taller bridges. Cliff Swallows prefer to nest on bridges made of concrete, as mud nests attach more securely and are protected by overhangs, and on those that span bodies of water since insects are plentiful in the air above water and swallows feed while in flight.

Bridges of intermediate height and those surrounded by a high diversity of habitats provide the greatest species diversity. The Barn Owl, for example, can be found nesting under concrete bridges surrounded by grasses and shrubs where rodents such as mice and voles may be found.

Ecotones, which are transition zones between two ecosystems, may be found between the under-bridge habitat and the surrounding habitat. Unmaintained or ignored sites encourage a diversity of species. Most ecosystems beneath bridges are highly stressed, with relatively few species, whereas the surrounding habitat may be diverse in comparison. The size and composition of the plant community are largely determined by disturbance levels (Bullock and Gregory 1991) and the amount of light available under the bridge. Soil quality in harsh habitats typically does not improve because of the limited establishment of early successional stages.

Landfills

Landfills are a common and relatively cheap form of waste disposal. A large city usually generates about one tonne of waste per resident annually, half is residential, originating from the home, and the other is commercial/industrial waste generated to support infrastructure and amenities for residents. Most of this waste is commonly buried underground in a landfill.

Landfills are usually located on the outskirts of cities, sometimes in areas of sensitive wildlife habitat. The major landfill of Greater Vancouver, BC, is in Burns Bog, a 40,000 hectare, environmentally sensitive area of which about 700 hectares is dedicated to the landfill.

Landfills have a number of environmental impacts besides the destruction of wildlife habitat, including leachate and production of gas. Landfills produce methane and sulphur compounds, which can be toxic to wildlife and humans.

Methane can be collected and burned. Restricting sulphur-bearing wastes, such as drywall, from the landfill, controls sulphur compounds.

Leachate is water leaving the landfill after rain and groundwater seeps through the garbage, picking up toxic chemicals or the products of decomposition from the waste. It then drains into the groundwater and eventually reaches streams, rivers, and lakes, where it contaminates these systems and poses a significant environmental problem. Capturing the flow of leachate with a liner or in a collection pond and removing or treating the waste can minimize the impacts.

Landfills attract scavenging animals, such as seagulls, rats, crows and bears, that feed on the garbage. Landfills encourage larger numbers of scavengers, which then have a negative impact on other wildlife. For example, crows may eat the eggs and fledglings of many songbirds that may live some distance from the landfill.

The chemical contaminants from a landfill may also be toxic to wildlife, either directly in small concentrations or in larger concentrations that result from bioaccumulation within an animal or biomagnification through the food chain.

There is no vegetation on the landfill site itself while active. Once the landfill is closed it is usually capped with sand and gravel. Escaping methane gas and toxic chemicals may prevent many pioneer species from becoming established. The site may be hydroseeded to provide a cover of grasses to prevent excessive soil erosion until it can be colonized by natural vegetation.

Rock Debris

Rock debris occurs in open areas of rock and gravel on sand or topsoil. These areas range in size from 2–3 m along roads to greater than 1 ha. Excess rock and gravel are usually pushed to the side after the completion of road construction. This space is graded and leveled to create roadside parking for vehicles. Vacant lots are often areas that are not yet developed or areas that have not been included in the development of the adjacent areas. Again excess rocks and gravel from construction on adjacent lots may be pushed into the vacant lots. In some cases, the rocks and gravel may

Rock debris can be as small as a pile of rocks in a yard or as large as a gravel pit. Although largely devoid of vegetation, they have their own unique faunal associations.

have been left in mounds and piles. This may create interesting niches for a variety of vegetation.

Moss and lichen often colonize bare rock. These plants excrete chemicals, which aid in degradation of the rocks and enable them to extract nutrients from the rocks. These activities contribute to the weathering of the rocks, which is the basis of soil production. The dead organic matter of the mosses and lichens contributes to the humus accumulation in soil. Other organic debris is brought into the area by wind and rain.

Other herbaceous plants and grasses that have been introduced to the area through seeds dispersed by wind and rain begin to take hold and germinate in the thin layer of soil. Many of the legume colonizers, such as clover and lupine, and trees such as alder, are capable of fixing atmospheric nitrogen, thus contributing to and increasing the levels of nitrogen in the nutrient-poor soil.

Bees frequent the patches of clover collecting nectar. Winged aphids invade young plant communities and form stable populations of the vegetation. As their population increases, their predators, such as the ladybugs, follow suit. The European Starling, Savannah Sparrow (*Passerculus sandwichensis*), White-crowned Sparrow (*Zonotrichia leucophrys*), House Sparrow and crows frequent these open areas. Domestic cats often visit these areas during their daily prowl.

Vacant lots may hold a surprising amount of insect biodiversity that undergo a succession of species composition.

Vacant Lots

Vacant lots with brick and concrete rubble are areas where buildings once stood. Once buildings have been demolished and removed from a site, broken concrete and asphalt slabs often remain.

Vacant lots may have some small rubble piles with concrete, brick, rock, and asphalt on topsoil. For example, an old school site may once have had a playground made of gravel. When left unattended and undisturbed over a period of time, colonizing species, both plants and animals, invade the site.

Vacant lots may be left undisturbed for a number of years prior to redevelopment. Plants and their associated wildlife quickly colonize these areas. The physical and chemical status of the area is often the factor restrictive to plant growth. Vacant lots usually have varying amounts of building material debris left on site. The size and amount of the debris are important factors in determining the extent of vegetation growth. Vacant lots' soils are usually dominated by building debris. Materials left on the site can create spaces and air pockets, which contribute to easy drainage of water and high oxygen content in the soil; both factors promote rooting of colonizing plants. The areas can eventually become highly productive ecosystems.

Nitrogen in the soils of the demolition sites is usually deficient and inadequate for plant growth. The common sources of nitrogen are wind blown organic debris (leaves, insects etc.) and debris brought in by rain. Accumulation of organic matter then functions as a nitrogen pool made available by the decomposition process. Nitrogen fixing plants, such as lupine and alder, are capable of fixing atmospheric nitrogen and making it available to the soil. Vegetation growth is dependent on the nutrient status of the soil. Soils in these areas are typically deficient in nitrogen, but have adequate phosphorus content (Bradshaw and Chadwick 1980).

Wildlife colonizing these areas are often mobile, such as insects and birds. The presence and abundance of wildlife is attributed to the amount of food at the site. In the initial seral stage, herbivorous winged insects invade these areas and utilize the colonizing plant as their source of food. Lazenby (1983, 1988) found that ground beetles (Carabidae) are among the first insects to colonize and establish their population in the rubble in the demolition sites. The changes of vegetation composition with succession are quickly followed by changes in wildlife diversity. Vole runs become observable in the grassland stage and the presence of birds is influenced by the availability of cover provided by trees and shrubs.

Brownfields

Special concern may need to be paid to manufacturing plants, industrial or commercial facilities that may have residual substances in the soil left over from the previous use of the site. They may contain hazardous substances, contaminants or pollutants. When these properties are being redeveloped or reused, it may be necessary to remove or treat the affected soil.

The United States Environmental Protection Agency initiated its Brownfields Program in 1995 to provide guidelines for the proper redevelopment and re-use of brownfields and to encourage brownfields projects.

Natural Services Provided by Barren Landscapes

Barren landscapes provide fewer natural services than other ecosystem types discussed in previous chapters. They do, however, provide limited wildlife habitat and enable a modest increase urban biodiversity.

11

Paved Landscapes

Paved landscapes are perceived as sterile and incapable of supporting wildlife. However, many mammals, birds, and invertebrates invade these harsh environments and adjust their habits. These opportunistic species may depend on people for their survival and are often regarded as urban pests once their population numbers become uncontrollable.

Buildings and Pavement

Where does urban wildlife live? Like humans, they also take refuge in buildings. Buildings are somewhat analogous to a cliff environment: both are harsh, but capable of providing excellent habitat for wildlife. Like the ledges of a cliff face, buildings and other structures have many crevices, openings, and designed holes or pockets, which can function as their shelter.

Basements, underground or partially underground areas of buildings with limited access to outside, are used by many wildlife that prefer dark and damp areas. Underground and covered parking lots are used as habitat by many urban bird species, as are industrial buildings, warehouses, and storage facilities: high-ceilinged buildings with ample hiding, nesting, roosting, and perching places. Ledges are analogous to the ledges of cliffs with platforms used for communal roosting and nesting. Smaller crevices and holes created by dilapidated rooftops, and spaces between roof tiles and buildings, also provide wildlife habitat.

City centres are capable of supporting higher breeding bird populations than other areas for some species (Dorney 1979). Such birds are highly opportunistic and are capable of exploiting the urban landscape to their benefit. The Rock Dove or Pigeon (*Columba livia*) favours buildings with ledges for communal roosting, and they use sheltered crevices and holes for their nests. They forage in commercial centres, food and beverage establishments, and open spaces frequented by people. Rock Doves are efficient scroungers, always at arm's length from their benefactors. They are also common along railway lines where they forage for grain dropped trains.

European Starlings are common and have stable populations across North America. They often flock in large numbers while foraging, roosting, and nesting. They nest in the urban trees and also on ledges and crevices of buildings. The Crested Myna (*Acridotheres cristatellus*), native to China, was introduced to Vancouver, British Columbia early in the twentieth century. Like the European Starling, it commonly nests on buildings as well as in urban trees.

House Sparrows are the most dependent on humans for their survival. They nest in the small cracks and crevices of buildings and occasionally in bird-boxes. They forage in city parks, sidewalks, and areas of human congregation.

Bats may roost in roofs and other cracks of buildings in urban areas. They are frequently seen at dusk flying from buildings that are near water. Other mammals common to building and paved areas include feral cats (*Felix domesticus*), dogs (*Canis familiaris*), house mice (*Mus musculus*), Norway rats (*Rattus rattus*), coyotes, striped skunks (*Mephitis mephitis*) and raccoons. These mammals are scavengers and are wildlife pests in cities. Norway rats originated in Europe and were accidentally transported to North America on cargo ships. They can be found in large numbers under wharves, docks, in the sewers systems and service ducts. They are voracious feeders and can cause serious damage to buildings. Surprisingly, raccoons, have adapted well to living in urban centres. They use storm drain systems and culverts as homes and passageways from one foraging site to another. Being omnivorous, they scavenge for food in garbage bins.

The undersides of roof tiles are good nesting sites for Starlings and House Sparrows, and rooftops are used by gulls, nighthawks and even Canada Geese, while small vents are used by Tree Swallows.

Dark and damp parts of buildings such as basements are ideal habitats for invertebrates. Cockroaches, a common household pest, are nocturnal and scavenge for food in garbage bins and kitchens. Silverfish (*Lepisma saccharina*) and firebrats (*Thermobia domestica*) feed mostly on starchy materials. They are attracted to starch in glues and laundry starch, causing serious damage in libraries, as well as linens, draperies and clothing. The common housefly (*Musca domestica*) is one of the commonest insects found all over the world. Houseflies multiply rapidly and can establish a viable population in a matter of days. They are carriers and transmitters of diseases.

Urban centres are landscaped beautifully with ornamental plants and trees, which are regularly maintained but are harsh environments. They have to cope with periods of extreme heat and drought. They are constantly exposed to pollution from exhaust emissions.

Some lichens will invade the concrete faces of buildings, roads and highway barriers. Various weeds are found growing in sidewalk and pavement cracks and in some open spaces. Some plantain and grass species colonize old gravel rooftops of buildings. In 1994 Maj Rope, a Milieu Planner, examined the vascular plants growing on a 1 km long section of a central street in Lahti, Mariankatu (Finland) and discovered that there were 59 species of vascular plants growing on the pavements (www.sci.fi/~evi10/kasvi1.htm#kasvit). Most common were *Plantago major*, *Poa annua, Matricaria matricarioides, Sagina procumbens* and *Capsella burse-pastoris*.

Containers can create habitat anywhere but generally require a lot of maintenance.

Container Habitat

People with little or no garden often do their gardening in containers. Gardeners and landscapers use containers to grow flowers, vegetables and herbs. Containers can be made of wood, clay, concrete, ceramics, and bricks, and vary in size and shape. Containers are common in open plazas, concourses, balconies, and lobbies of buildings, hotels, and apartments. Containers can also be used to accentuate the beauty of a landscaped residence.

Soil used in containers is usually a mixture of compost, washed sand, sterilized garden topsoil, and peat moss or bark mulch. Fertilizers are often added before planting and are required throughout the growing season

Petunias, fuchsias, geraniums, and lobelias, as well as ivy, are commonly grown in hanging baskets, which are very attractive to hummingbirds. Seasonal bulbs add beautiful colours as each season progresses. Marigolds do exceptionally well in containers, adding colour under shrubs and dwarf trees.

Native shrubs, such as red osier dogwood (*Cornus stolonifera*), salal, and Oregon grape, are also commonly used not only because of their easy care and maintenance but also because of their aesthetically pleasing growth habits.

Birds are usually attracted to the different colours of flowers. Butterflies are especially attracted to sweet smelling flowers that advertise the presence of nectar. Wasps and bees are common pollinators of members of the Family Compositae, such as chrysanthemums and daisies. Spiders spin their webs in the small branches of shrubs and trees, trapping insects in flight.

Alleys and Lanes

Alleys and lanes are a variable habitat. They can be stark, with little vegetation. The ground is compacted or paved, and there are frequent disturbances from cars. In residential areas, lanes are often lined with grass and other vegetation, and disturbances from cars are less than in city centres. These are linear environments that act as corridors for the movement of wildlife through the city.

Several distinct categories of alleys and lanes can be distinguished to reflect the flora and fauna, which might occur in each. Downtown alleys are paved and surrounded by buildings. There is no natural vegetation except for the occasional tuft of grass or patch of weeds growing in cracks. The only source of food for wildlife would come from garbage.

A nest of a Glaucous-winged Gull in a rooftop container. Robins and House Finches will nest in flower pots or hanging baskets — one House Finch nested in a hanging basket on the 26th floor of an apartment building.

Residential alleys are found in most residential areas. They may or may not be paved. Garages may back onto the alley and garbage cans are frequently stored along its borders. Vehicle traffic is sporadic and not as heavy as in downtown alleys. Backyards line the edges and are frequently planted with shrubs and trees.

Lanes occur in less densely populated areas. They are characterized by having little vehicle traffic and they are usually not paved. They are more natural than alleys and are usually distinguished from the adjoining farmland and green space by fences.

Alleys and lanes, along with natural corridors such as streams, ravines, and artificial corridors such as railway lines, provide a continuous network of passage for wildlife to travel though the city. Coyotes, skunks, and raccoons regularly use them to gain access to foraging sites in the heart of business and commercial districts downtown. Numerous perches and hole-nesting or cliff-like nesting sites provided by buildings, utility lines, and garages attract many urban bird species such as pigeons, European Starlings, House Sparrows, and House Finches (*Carpodacus mexicanus*).

The residential alleys attract feeding flocks of Black-capped Chickadees, Bushtits (*Psaltriparus minimus*), and Dark-eyed Juncos (*Junco hyemalis*), which enjoy foraging on the trees and shrubs along these edge environments. Northern Flickers (*Colaptes auratus*) are also found in this habitat. People frequently plant berry-producing plants that attract Cedar Waxwings (*Bombycilla cedrorum*) and American Robins (*Turdus migratorius*). If the alley is close to a large open field or a larger urban wilderness park, black-tailed deer (*Odocoileus hemonius*) or black bear (*Ursus americanus*) may also wander through.

Alleys and lanes may also attract less desirable species. Rats and mice are usually found here in association with the garbage and backyard composts, and the many hiding places provided by woodpiles and garages.

Plazas

Plazas occur throughout cities in public areas. Plazas can be found on street corners, next to cathedrals and institutional properties, and outside malls, transit stations, and libraries.

They resemble lush pockets of vegetation and may contain pools or fountains. They are often built slightly sunken or risen away from busy streets, which reduces motor vehicle disturbance for the public. Plaza design often takes advantage of current land use and conditions. Plazas surrounding transit stations often function as important natural green space for wildlife and people.

Plazas typically consist of a variety of native and non-native plants. Ornamental shrubs are very common and, generally, define the outline of a garden setting, giving texture, form, and colour in countless ways. To create a more natural appearance along with a wide variety of ground cover, lush trees of many kinds are selected. Some examples are Japanese cherry (*Prunus sp.*), honey locust (*Gleditsia triacanthos*), and ornamental magnolias *(Magnolia sp.)*. Soil varies from compact to

Plazas offer great potential to create habitat, especially if they incorporate a fountain.

loose and in some areas it is covered with bark mulch. The general growth forms include woody ground covers, sedges, climbing vines, grasses, and trees.

Plants are selected to satisfy the environmental and cultural demands of urban conditions. Often, plants must be cloned and grafted to withstand environmental requirements. They must be resistant to a growing number of diseases and to leaf rust and insects. They must withstand drought and restricted soil conditions and leaves must be tolerant of urban air pollution.

Plant density within a plaza varies depending on the location, vegetation, sun exposure, and maintenance. As the diversity of vegetation increases, so does ecosystem function and balance.

Plazas provide a link to the animal world. They support many different species of wildlife and offer sources of food, places to live, and nesting sites throughout our scattered green spaces. Some of these are birds, mice and rats. Birds that seek refuge in downtown plazas include the European Starling, House Sparrow, and Rock Dove, which are a few of the most abundant species in the city. Plazas are also used by transient species, which use the plaza as a resting spot while moving around and between cities. Plazas with open areas of grass may attract certain types of birds, like the Canada Goose and American Robin. If the plaza contains a pool of water, this might also attract water birds, such as Mallards.

Natural Services Provided by Paved Landscapes

As with barren landscapes, the natural services provided by paved landscapes are limited. They include sequestration of carbon and removal of other pollutants from the atmosphere and wildlife habitat.

12

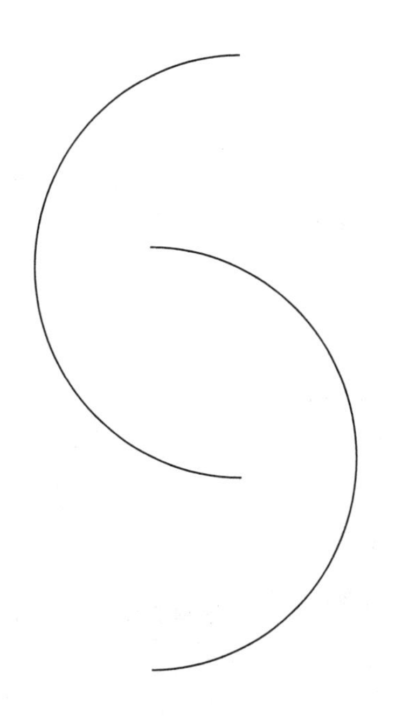

Corridors

Corridors

Corridors are linear areas that allow passage of utilities, trains, and other traffic. These corridors are abundant in urban areas and provide an opportunity for wildlife to move throughout the city.

Utility Corridors

Utility corridors are undeveloped linear plots that are necessary to allow safe passage of power lines, gas pipelines, water mains, and sewer. Due to low maintenance, utility corridors can be overgrown with shrubs. Power corridors are easily identified by tall metal structures with large wires connecting them. The high-tension wires are often accompanied by a low humming sound from the electricity. Gas and sewer corridors tend to be open land with the piping buried beneath.

Utility corridors offer an excellent opportunity to salvage young trees for planting elsewhere. They would otherwise have to be cut by the utility company to keep them from posing a hazard to the overhead powerlines.

Utility corridors begin at the source of the utility, such as power, natural gas, or sewage treatment plants. The larger transmission corridors radiate from this point. Smaller distribution corridors often branch from the main corridors. The main corridors can be found traversing most urban and rural habitats, including mountains, valleys, and rivers. The smaller branches are more common in urban areas, where they carry the utility to neighbourhoods.

Utility corridors transect many urban habitats and the maintenance level of these corridors is dependent on their type and location. Buried corridors are frequently mowed to ensure that pipes can easily be maintained. Power corridors are usually minimally maintained, except where they intersect developed urban areas. These corridors are frequently used as part of an urban trails system. Grasses, herbs, and shrubs colonize these maintained areas. The soil quality and type will resemble the soil of the related habitat.

Species diversity is high on most rights-of-ways because of the edge effect, a phenomenon caused when more than one habitat type come together. Species from both habitats invade from the edges and colonize the middle ground. This leads to a high number of species. The corridor is also important as an urban path between isolated habitats, allowing separated populations of terrestrial animals to breed, increasing the genetic diversity. A linear habitat also allows animals to have a larger home range. A block habitat may have the same area, but movement is restricted compared to a long, straight utility corridor that may connect to other green spaces.

Utility corridors offer an opportunity for community groups to plant shrubs and perennials to create habitats that enhance connectivity, helping wildlife, especially birds, move between parks and other urban green spaces.

Maintained corridors are often planted with lawn grasses like fescue and blue grass. Species like creeping buttercup, clover and blackberry runners can also be found. Mowing tolerant plants like plantain (*Plantago sp.*) and common dandelion (*Taraxacum officinale*) may also be present. In contrast, the minimally maintained corridor will have wild grasses like canary grass and cut grass, as well as larger shrubs such as hardhack and blackberry thickets. Alder, black cottonwood, and various conifers will also begin growth here until maintenance removes them. Finally, an unmaintained corridor will gradually move through various successional stages, with grasses giving way to shrubs and saplings until a climax forest may generate if conditions allow.

Animal species present on the maintained corridors will include small birds like Robins and Starlings and, possibly, small mammals like voles. Minimally maintained corridors will have more birds present, such as Red-winged Blackbird, Ring-necked Pheasant, and possibly smaller raptors like the Northern Harrier. Mammals like coyotes and raccoons may take up residence if the cover is thick enough. Larger populations of vole, mice and rats are likely.

Greenways

Greenways are an important element of modern urban design. They add much to the livability of a city. They are seen as important in protecting natural areas and in providing recreational opportunities. They can range in size from a narrow urban trail to a wide natural landscape linkage. The term greenway is relatively new, appearing in the 1950s, but the potential of greenways to provide access to city parks was recognized as early as the 1860s by Frederick Olmsted who proposed such linkages between Berkeley and Oakland, California, and later in New York City in the design of Brooklyn's Prospect Park (Smith 1993).

Greenways have been variously called urban trails, landscape linkages, greenbelts and ecological corridors, reflecting differences in their structure and function. Whatever the case, an integral part of a greenway is often a trail for people. Nevertheless, although greenways are inherently for people, they also can be of great ecological value if they also have the vegetation to serve wildlife.

Greenways encourage interbreeding among plant and animal populations. The corridors provide opportunities for animals in particular to move from one green space, such as a park, to another. Populations from these different fragments of green space have a better opportunity to interbreed, forming a larger metapopulation. In addition to increasing the likelihood of finding a breeding partner in the first place, this adds to the genetic diversity of the species, improving resistance to disease and improving the resilience of the species.

Greenways have evolved over the years. Searns (1995) identifies three generations: Generation 1 greenways are the boulevards and parkways of the green city. Generation 2 greenways are trail-oriented, providing access to rivers, streams and other corridors. Generation 3 greenways, though, are multi-objective; they go beyond recreation and beautification and incorporate wildlife needs and other ecosystem functions. The level 3 greenway is focused on biodiversity.

The greenway creates wildlife habitat as well. Many greenways incorporate places for stormwater retention, creating temporary wetlands. The vegetation in the greenway can serve as a filter for sediments and pollutants, preventing them from entering healthy streams and lakes. The increased mobility of species between green spaces results in more dynamic and stable ecosystems. An infinitely complex series of interactions can occur, establishing and maintaining healthier ecosystems that can better accommodate disturbances impinging upon them from surrounding urban areas. There is better water quality, more stable water cycles, healthier habitats with more diverse flora and fauna, and more meaningful connections to nature for people within a community. Trees in greenways act as windbreaks, reducing soil erosion.

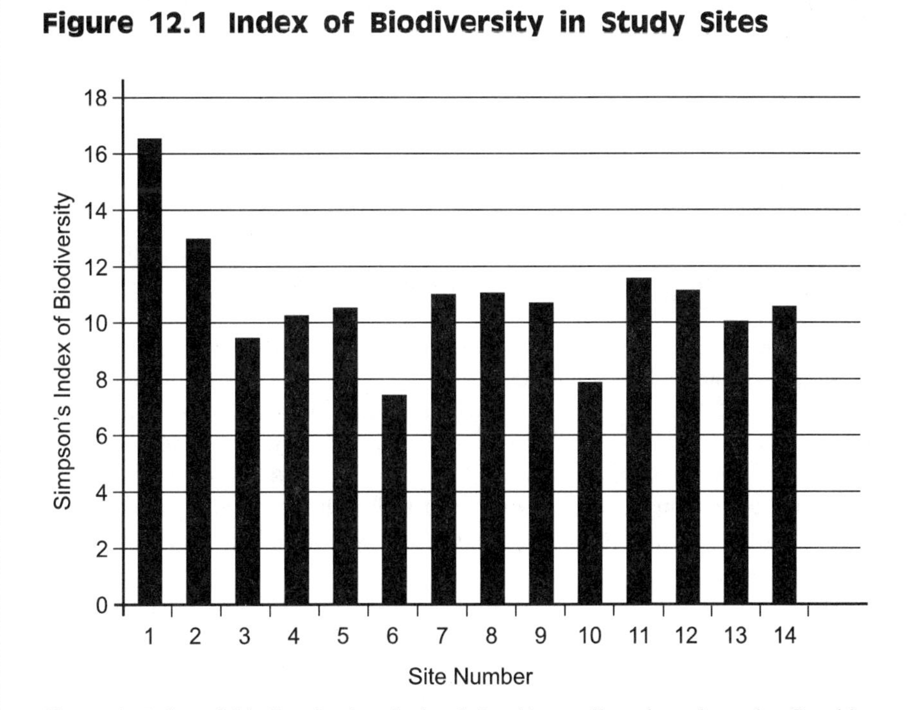

Figure 12.1 Index of Biodiversity in Study Sites

Simpson's Index of biodiversity is calculated for 14 sampling sites along the Coquitlam corridor. Site 1 is the farthest north, next to the wilderness fringe. The five patches of green space (environmentally sensitive areas) joined by the utility corridor are represented by Sites 4, 5 and 6 adjacent to Scott Creek Ravine, Sites 7 and 8 adjacent to Pinnacle Creek Ravine, Site 10 adjacent to Mundy Park, Site 11 adjacent to the Riverview Lands and Site 14 in Colony Farm Regional Park (Schaefer and Sulek 1997, Vol. 1).

Although greenways can provide habitat, it may not be as good as that available in the parks and other patches of green space by the greenway. Schaefer and Sulek (1997) found that the diversity of songbirds in a greenway next to a large green space is less than the biodiversity in other parts of a greenway not in the vicinity of a large green space. It appears that, if a choice is available between the habitat of a greenway versus that of a large green space, the birds choose the large green space.

Establishing greenway networks that benefit both people and wildlife is challenging. In many cases, cities have limited natural open spaces with which to plan greenways. In a comparison of Milwaukee, Wisconsin and Ottawa, Ontario, Erickson (2004) found that collaboration between the diverse government departments and public interest groups involved in greenway planning and implementation is piecemeal. Other challenges include changing objectives for greenways over time and a lack of a coordinated greenway vision.

Greenways can provide numerous economic benefits to a community as well. The extra green space and recreational opportunities a greenway can offer attract more economic development. Houses backing onto greenways generally have higher property values — one studied identified an increase of 11–15% (Quayle and Hamilton 1999). Greenways can also serve as green belts to manage urban sprawl.

Ravines

It is difficult to decide if a ravine is a corridor, a stream, a riparian habitat or a forest. Ravines are valuable habitat to numerous species who live and breed in these often secluded environments.

A study of urban ravines in Greater Vancouver, British Columbia (Schaefer et al. 1992) documented 140 ravines with a total area of over 15 km^2 within urban limits. About one third of the ravines occur in parks. The remainder are mainly in private or municipal land (and therefore may be destroyed by development), with a few being a mixture of park land and some other category. The ravines occur along the escarpments of the Burrard Peninsula, the Fraser River, and Boundary and Mud Bays. A few also occur on the flatlands along major creeks.

Some of the ravines in the Greater Vancouver study were only selectively logged in the past and retain few of the original trees. Several of the ravines contain stands of very large Douglas-fir, Western redcedar, and Western hemlock. Individual examples greater than 0.7 metres in diameter are common, and

Although ravines represent some of the best natural habitat in densely urbanized areas, they are frequently neglected and used as garbage dumps. A ravine clean-up is a first step to restoring these important natural areas.

some were more than one metre in diameter, despite the fact that just over the tops of the banks might be a thriving, high density urban community.

Boulevards and Street Trees

Boulevards occur along many major urban roads. They can be several metres wide, affording enough habitat to support a high diversity of vegetation. These areas are usually well-maintained and are without a well-defined understorey.

Street trees attract a large variety of birds and mammals. They act as linear corridors for warblers, chickadees and other canopy feeders to pass through the city to surrounding forests.

Street trees can reach considerable size and age. Such older trees are often native veteran trees that were once part of the original ecosystem in the area and have somehow managed to survive development. They may also be non-native heritage trees that were planted long ago to commemorate some anniversary or event, or were planted by someone of importance in the community.

Based on a study of street trees in Davis, California, Maco et al. (2002) suggested that 25% coverage of streets and sidewalks was possible in view of land use, planting locations and age distribution. However, they point out that most cities do not have a performance ordinance for street trees that requires a certain percentage of canopy cover, as they might for other paved areas.

The monkey-puzzle tree dates back to the age of dinosaurs. It has hard, pointed, scale-like leaves with long tail-like branches.

For example, an ordinance adopted by Sacremento, California in 1983 requires parking lots be covered with 50% shade within 15 years of development. Also, it is a requirement on residential lots in many cities to have one tree every 10–20 m of street frontage (Abbey 1998).

Shading of paved areas increases pavement longevity (McPherson et al. 1998), helps to control stormwater runoff (Xiao et al. 1998), mitigates urban heat island effects (Akbari et al. 1992) and helps to reduce hydrocarbon emissions involved in forming ground level ozone (Scott et al. 1999).

Boulevard trees are usually a mix of native and introduced species. A study (Pinnell et al. 2004) of street trees in Coquitlam, BC, determined that 51% of the trees were non-native (Figure 12.2). Municipalities often prefer to plant at-

Figure 12.2 Proportion of Native/Introduced Species Among Street Trees in Coquitlam

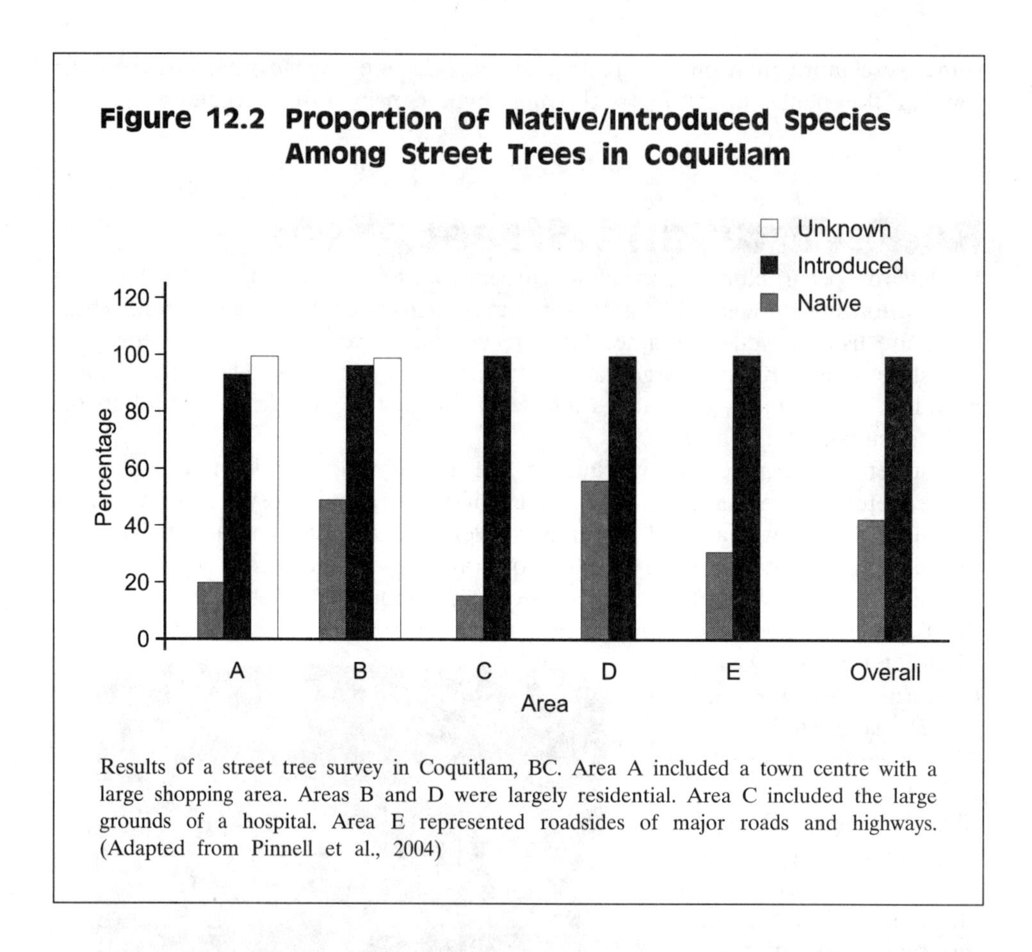

Results of a street tree survey in Coquitlam, BC. Area A included a town centre with a large shopping area. Areas B and D were largely residential. Area C included the large grounds of a hospital. Area E represented roadsides of major roads and highways. (Adapted from Pinnell et al., 2004)

tractive species with showy blossoms or species from eastern North America that occur in cities and are therefore known to be tolerant of the increased pollution. Common boulevard species are also of Mediterranean origin and are better adapted to the drier conditions of an urban environment.

Collectively, boulevards and street trees create an atmosphere of an open forest ecosystem. Tree varieties have been bred to suit the needs of the urban environment and can be tall or short, spreading or narrow, colourful, or non-fruit bearing (a hazard for cars on roads). Deodar cedar, a true cedar native to the Mediterranean, has needles instead of scales. This tree grows to be large and lush, adding a beautiful texture to the urban landscape. Several varieties of European beech are also common. The copper variety forms a large canopy with a thick trunk and has a stunning reddish colour. A different cultivar has green leaves and grows straight and tall, so it can be planted next to a building.

The urban forest of boulevards also regularly contains London plane trees, related to the sycamores of eastern North America. They have a characteristic peeling patchwork on the bark and a maple-leaf shape to the leaf. Tulip trees (*Liriodendron tulipifera*), native to eastern North America and resistant to pollution, are frequently planted with them. Other broad leaf trees seen along roads and in people's yards include Lombardy poplar (*Populus nigra*), white poplar (*P.*

alba), common horse chestnut (*Aesculus hippocastanum*), ornamental cherry and plum, green ash (*Fraxinus pennsylvanica*), northern red oak (*Quercus borealis*) and English oak (*Q. robur*). Conifers often seen as street trees include Lawson cypress (*Chamaecyparis lawsoniana*), which has many cultivars varying in size and shape from a low shrub to a tall tree. Junipers (*Juniperus sp.*) and giant sequoia (*Sequoiadendron giganteum*) are common as well. There are also the really unusual looking trees such as the monkey-puzzle tree (*Araucaria araucana*) native to Argentina and Chile. It is related to the Norfolk Island pine (*Araucaria heterophylla)*, which is a common houseplant.

Street trees next to sidewalks or areas used by dog walkers sometimes demonstrate a "canine zone" where their appearance is changed because they are frequently urinated on by dogs. The zone may appear dark in cities with moderate air pollution. In Europe, the trunk of such a tree normally appears pale grey-green because of the lichen *Lecanora conizaeoides*, or bright green because of the algae *Pleurococcus viridis*. However, when frequently "visited" by dogs, the tree will develop an epiphytic flora of the algae *Prasiola crispa*, *Hormidium flaccidum* and *Strichococcus* sp., along with the lichen *Lecanora dispersa* and the moss *Ceraton purpureus* (Gilbert 1991).

Inappropriately planted street strees may need to be heavily pruned later in life to avoid damage to overhead wires. This may seriously deteriorate the health and longevity of the trees and their importance to biodiversity.

Cities are sometimes reluctant to plant boulevards because of liability concerns and the cost of maintenance. An example of having residents assume stewardship of boulevard plantings is the Blooming Boulevards project in Vancouver, British Columbia (James 2003). The city worked with residents to plant boulevards between the sidewalk and road in front of their houses. Residents were offered small grants for plant material and supported plantings that met the following guidelines:

1. All plant material should be less than 1 m in height to preserve sightlines for drivers.
2. Some plants must provide structure for winter interest.
3. Gardens were to be mounded 20–30 cm to allow for root growth and to further separate the plants from below-ground services.
4. Parks department arborists were to be consulted for plantings around trees and roots.

Table 12.1 Drought-Tolerant Plants

Some of the drought tolerant perennials recommended by Van Dusen Botanical Gardens for the Blooming Boulevards Project were:

Botanical Name	Common Name	Sun	Shade	Bloom Season	Height
Achillea millefolium	Yarrow	×		Summer/fall	1 m
Aethionema sp.	Stonecress	×		Spring/summer	15 cm
Alchemilla mollis	Ladies Mantle	×	×	Summer	30 cm
Allium	Chives	×		Summer	30 cm
Anthemis nobilis	Chamomile	×		Summer	60 cm
Aqueligia sp.	Columbine	×		Summer	30 cm
Berengia cordifolia	Elephants ears	×	×	Summer/fall	20 cm
Carex sp.	Sedge	×	×	Spring	30 cm
Centhranthus sp	Valerian	×	×	Summer	60 cm
Ceratostigma sp.	Plumbago	×	×	Late summer/fall	30 cm
Deschampia sp.	Tufted Hair Grass	×	×	Spring/summer	30 cm
Dryopteris sp.	Wood Fern		×	N/A	75 cm
Echinacea purpurea	Cone flower	×		Summer	1 m
Erica sp.	Heather	×		Winter/spring	30 cm
Erigeron sp.	Fleabane beach aster	×		Summer	20 cm
Erysimum sp.	Wallflower	×		Summer	30 cm
Frageria chiloensis	Wild strawberry	×	×	Spring/summer	30 cm
Geranium sp.	Geranium	×	×	Spring/summer	30 cm
Helictotrichon sp.	Blue oat grass	×		Spring/summer	30 cm
Hemerocallis sp.	Daylily	×		Summer	30 cm
Iris sp.	Irises	×	×	Summer	1 m
Lavendula sp.	Lavender	×		Summer	60 cm
Lupinus sp.	Lupine	×		Spring/summer	30 cm
Papaver sp.	Poppy	×		Spring/summer	45 cm
Pennisteum alopecuroides	Fountain grass	×		Spring/summer	1 m
Salvia sp.	Sage	×		Spring/summer	60 cm
Sedum sp.	Stonecrop	×	×	Spring/summer	20 cm
Senecio cineraria	Dusty miller	×		Summer	30 cm
Stachys byzantina	Lam's ears	×	×	Summer	30 cm
Thymus sp.	Thyme	×		Summer	60 cm

Linear Riparian Areas

Linear riparian habitats of grass and shrub border on urban waterways and exhibit a high degree of variability. Differences in physical appearance are due to variables such as soil moisture, fertility, disturbance, and competition. One of the most important factors determining the type of linear environment bordering the waterway is the size and type of water channel.

Riparian zones of rivers and large streams tend to have larger transition zones between the water's edge and the vegetation lining the waterway above the maximum flood level. This larger area gives rise to a distinct set of sub-zones capable of accommodating a greater diversity and plant species.

Smaller streams and drainage ditches may not have a gradual well-defined transition zone between the water's edge and vegetation. Often, as is the case with most drainage ditches, the change is very abrupt. The distance between the water's edge and the grass bank may only be a few, almost vertical, feet, and thus the habitat's capacity to support a greater diversity of species is reduced.

Riparian habitats occur along the banks of almost every type of waterway flowing through or within urban environments. In large urban centres these habitats are often part of a linear park system and are usually subject to heavy pressure from recreational use and from management by park maintenance staff.

The linear habitat along rivers, streams or ditches is often composed of three distinct sub-zones. The extent of differentiation into the different zones is directly related to the size and type of water channel. There is a great degree of variability in the complexity of the various waterside habitats. Mudflat or foreshore areas are subjected to periodic cycles of flooding and dryness due to tidal effects, variations in flow rates due to precipitation run off, or flood- or drainage-control programs.

Riparian zones are popular areas for habitat restoration because they support a variety of plant and animal life.

There is a transition zone between the water channel and the maximum flood height. This area often lies on a moderate to steep gradient, which rises from the water's edge and is not subject to regular flooding. In the transition zone between the water channel and the maximum flood height the soil is often of the same composition as the vegetated bank area above it, but with increasing amounts of sand, silt and mud as you approach the water's edge. Because this portion of the bank usually lies on a gradient leading down to the water, it is more susceptible to soil erosion, landslides, and flooding in the lower portions. Frequently on larger waterways, embankments consisting of large amounts of scree and talus, which are formed into dykes, may be incorporated into this area as a means of flood control.

Riparian zones of rivers and large streams attract larger numbers of wildlife species and, because of their more gradual ecotones, they also support a greater diversity of plant life. Some typical examples of wildlife include various waterfowl, such as Canada Geese or Great Blue Herons in the foreshore areas, and frogs, raccoons, moles and various songbirds in the other areas.

Typical plants such as the grasses, clover (*Trifolium spp.*), berry bushes and shrubs provide abundant cover, nesting, and perching sites, along with food and forage, to support wildlife. A few trees in forested riparian zones, such as the black cottonwood and birch, which favour damp and wet conditions, are also characteristic of these locations.

Smaller waterways do not exhibit as much species diversity because of the smaller habitat areas. Some typical wildlife examples from these areas may include muskrats, frogs, foraging ducks, and raccoons. The plant species associated with these areas are almost always limited to grass species and smaller shrubs. Horsetails and ferns are also typical examples of other water loving vegetation found in these areas.

Railways

Railroads are used for the transportation of industrial, agricultural, and resource products and can be found in areas where these products are manufactured, grown, harvested, shipped, and delivered. These habitats are linear and continue for hundreds of miles. The two major forms of track found in urban areas are active tracks and yards. Most tracks in urban areas are used by transfer trains and are found in industrial areas. Yards are collection centres for trains and can be found in central locations throughout urban centres.

The main difference between transfer track and rail yards, from a habitat point of view, is the amount of switching. Switching involves violently crashing cars together to couple them. This crash often knocks grain and other plant seeds loose from the car's undercarriage. These seeds cause the habitat difference. Trains travel great distances across the continent and pick up various seeds along the way. When these cars enter a yard, many of these seeds from afar fall and take root in the yard. This seeding of foreign plants is not as common on main and transfer tracks.

Plants that grow on railway rights-of-way are short lived. Most foreign plants do not propagate (Gilbert 1991) and those that do, along with native plants, are killed by maintenance crews using either herbicides or steam. The plants are destroyed to prevent soil formation on the right-of-way because soft soil destabilizes the tracks.

There is no soil on the rail bed; any that is formed is actively removed to prevent track destabilization. The rail bed is made of a layer of cinder under a heavy layer of large gravel. The gravel is up to two feet deep, with sloping sides.

A distinction must be made between the animal species that inhabit the track bed and those that use the tracks opportunistically to feed or move. No large vertebrates can inhabit the track because of the frequent maintenance and lack of soil and only the hardiest invertebrate species can survive, feeding mainly on grain spillage. Grain spillage does attract a variety of large animals to feed on the tracks. Canada Geese can be especially bold, even entering railway maintenance shops in search of spilled wheat. As well, Rock Doves, Ring-necked Pheasants, rats and mice risk the tracks in search of food. In winter, large mammals like deer and moose (*Alces alces*) (in smaller cities surrounded by wilderness), are common on tracks, using them as a trail because deep snow is cleared from the tracks creating a snow-free path. In urban areas, railway tracks act as trails for mammals between isolated habitats, possibly enhancing genetic diversity and certainly allowing access to otherwise inaccessible areas.

A railway corridor through a city can be an amazing green space.

Plants that can tolerate the conditions on the tracks must either be adapted to being run over regularly or live well outside the tracks. The clearance between the rail bed and the train is as low as four inches at the locomotive's front, meaning that any plant that survives on the tracks must either be shorter than four inches or extremely flexible. Some of the plants that can survive here are plantain, fireweed, vetch, clover, creeping buttercup, dandelion and St. Johns-wort (*Hypericum formosum*). Outside the tracks, on the rail bed, verge alder saplings can start along with blackberry runners and goldenrod. Most plants here do not reach maturity and very rarely do they reproduce.

Corridors provide many of nature's services in urban areas. These services are very similar to those provided by urban forests and shrub communities. Due to the high amount of vegetation in many corridor types, sequestration of carbon and removal other pollutants from the atmosphere are important services provided by this landscape type. Corridors also provide wildlife habitat and work to improve genetic and biological diversity.

13

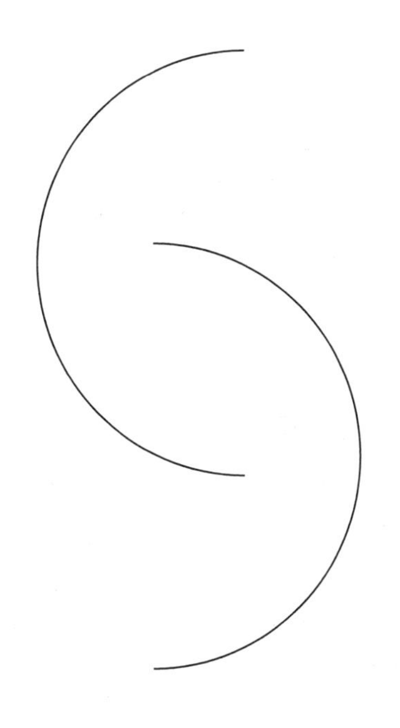

Garden Landscapes

Garden

landscapes are found in and around residential yards, parks, and businesses. These landscapes are often on privately owned lands, as is the case with residences, but can also be part of the public landscape, such as with parks.

Lawns and Yards

Lawns occur in residential areas and manicured parks for landscaping and playing surfaces. They typically consist of only a few species of grass, mainly fescues and bluegrass (*Poa sp.*), that are regularly mowed. The soil is generally compact because of high use. There is no cover for wildlife. In the United States it is estimated that lawns cover 2% (8 million ha) of the land area. Thirty plant and 100 insect species are commonly found in urban lawns (Garber 1987). Lawns in residential areas typically take up approximately one third of a house lot (Schaefer and Sulek 1997, Vol. 4). In standard city parks, the lawn area is about 75% (Gilbert 1991).

Lawns found around residences cover an average of about 30% of the total lot. They are also found around schools, public institutions, businesses, and playing fields and golf courses. Lawns tend to be more extensive away from the city core. Lawns frequently are the only type of open space found in the central core.

Lawns are generally flat, with no standing water and good drainage. They can be highly fragmented lawns in urban neighbourhoods where driveways cut across the lawn for road access, or they are intersected by gardens, hedges, or fences. Fragmentation can restrict the movement of smaller species. Continuous, mixed management lawns are where residential yards meet. Playing fields and parks lawns represent continuous, uniformly maintained lawns. These lawns are usually larger than two acres and owned by the city, although some more affluent neighbourhoods may have continuous, private lawns this large.

When clover is mixed with grasses in lawns, various nectar feeding insects, including bees and butterflies, are attracted to the lawn. Birds such as European Starlings, American Robins and others forage lawns for worms, larval insects, and flying insects. The presence of any perching sites or cover through landscaping or buildings greatly increases the diversity and abundance of wildlife. The presence of water near the site will attract ducks and geese, which feed on grass. Lawns contain a variety of soil invertebrates that attract birds and moles (such as the coast mole *Scapanus orarius*) in particular, large earthworms (*Lumbricus terrestris*) and *Tipulid* (crane fly) larvae. The species diversity here is low, but the biomass found can be surprisingly high.

Other plants often considered weeds, such as the creeping buttercup, plantain, and dandelions, are also found in lawns. This increases the availability of nectar and attracts a wider variety of transient insects. White clover (*Trifolium repens*) fixes nitrogen, naturally fertilizing the lawn. Weeds survive by adapting to the lawn environment. Dandelions have very deep roots that are difficult to remove and can reproduce through rhizomes. Plantain have adapted their leaf growth form to allow them to lie flat, avoiding the blades of the lawnmower. Native plants such as the creeping buttercup and trailing yellow violets (*Viola sempervirens*) can invade lawns using runners. As well, poorly drained, acidic, shaded sites tend to encourage mosses. Less management means more species.

Yards contribute to biodiversity not only through their vegetation, but because they allow water to percolate into the ground, feeding wetlands and streams in their watershed.

In the Pacific Northwest — southwestern British Columbia, Washington, Oregon and northern California — the mole is a common, albeit unwelcome resident of neighbourhood lawns. The two species of moles found in this area are the coast mole (*Scapanus orarius*) and the Townsend's mole (*S. townsendii*). Both are territorial, constructing molehills in pastures, crop fields, golf courses, cemeteries and yards. They are examples of urban exploiters and have an interesting history related to urban development. Their diet consists largely of earthworms — about 93% in the case of the coast mole (Glendenning 1959) — in particular the large night crawler *Lumbricus terrestris* that loves our lawns. However, this has not always been the case. Some 94% of the earthworm species presently found in North America, including *L. terrestris*, were introduced from Europe (Lindroth 1957), mainly in Earth used as ballast in ships coming without cargo from Europe centuries ago to bring back goods from America.

The worms spread across land from the east, with the help of settlers, and have only been in the Pacific northwest for 100–200 years. Although the diet of the coast mole is now mainly earthworms, its dentition, with flattened and rodentiform first incisors, would suggest that it used to primarily eat insects. The European mole (*Talpa europaea*), that has traditionally fed on earthworms all along, has a dentition typical of most carnivores consisting of large canines with small central incisors. It appears that the coast mole's diet is an example of prey switching (Schaefer 1984). Coast moles began to eat more earthworms when they became abundant due to agricultural and urban development.

Climbing Plants

The density of plant growth on building walls varies as a result of location, species composition, sun exposure, and maintenance intensity. The colour and density of plants also varies, depending on season, flowering potential, and type of vegetation.

Some mole species are urban exploiters. The coast and Townsend's moles benefit from lawns and agricultural lands that provide habitat for their preferred food of large earthworms. Both species are territorial and fossorial, living almost entirely underground. We usually just see their hills of dirt.

Climbing vegetation on buildings is becoming increasingly rare in modern urban settings. It may be difficult to locate outstanding examples of this type of plant community. In some older urban areas, or the older sections of evolving urban centres, a few buildings may exist that still sport a complete blanket of vegetative growth on their exterior structure. More common, however, are structures that are partially covered, such as concrete walls or chimneys. One factor contributing to the rarity of urban structures clothed in vegetation is the prohibitive cost associated with the management and upkeep of these vertical gardens.

Climbing plants comprise two types: shrubs and vines. While the climbing potential of vines is obvious, some shrubs can also be pruned and managed, causing them to climb like vines.

Climbers have a number of different applications, ranging from "roofing", framing doorways and windows, to curtaining walls. They can also be used to soften building lines, clothe pillars with living green, transform concrete walls, and lend colour to dull fences (Lenanton 1980).

English ivy (*Hedera helix*) is capable of attaching itself directly to walls, wood posts and masonry by means of modified roots called holdfasts. Since ivy does not require any special structural support, it can be found on structures lacking extensive surface texture or support. Other types of climbing species, such as *Clematis* or climbing rose, require varying degrees of structural accommodation to allow them to fully express their climbing potential. Fine support structures such as wires, small diameter runners, or stems from other vegetation will allow for *Clematis*, which twines its leaf stems around these supports to climb and spread along a structure. Other heavier plants, such as climbing rose and Wisteria will require stronger and more substantive support to accommodate their greater weight and prevent collapse under the strain of mature growth. These climbers must be trained to grow onto those supports.

Climbing plants can create a large amount of leaf volume with little or no footprint on the ground.

Some vines and shrubs can be manicured to grow in distinctive patterns against walls, fences and other structural supports. These are the espaliers that are used to create decorative shapes and designs, save space, and provide growth in narrow places where there is insufficient room for larger plants. An examples is firethorn (*Pyracantha sp.*) (Lenanton 1980).

The majority of climbing shrubs and vines in cities have a wide tolerance for various soil conditions. Many of the urban ornamental shrubs and climbing vines prefer slightly acidic, well-drained soils with lots of sun. Some climbers, such as ivy, have rootlets at every joint, which are termed aerial roots or adventitious roots. These rootlets are used mainly for support, but are capable of absorbing moisture and nutrients not available in the soil.

A well-established community of climbing plants has the potential to support a limited variety of birds. The density and abundance of growth and management regimes are primary factors in determining the capacity of this type of habitat to accommodate any wildlife. Structures with well-developed growth have good cover, nesting, and perching site potential for many songbird species such as House Finches, Sparrows, and Robins. Since the higher areas on walls, along rooflines (overhangs), and eaves troughs afford more protection from exposure and disturbance, these areas are highly prized urban bird habitats. The wildlife value of these habitats also increases greatly if the climbing plant species can also provide a food source, such as berries or nectar producing flowers. Butterflies and other insects attracted to the nectar may then become food for the resident bird community. Sites with thinner growth and more management in the form of frequent manicuring will not be as attractive to the many species because of the corresponding thin cover and frequent disturbance.

Although not usually native, ornamental plants produce a great deal of biodiversity.

Ornamental Plants and Shrubs

Ornamental horticulture depends upon a wide variety of shrubs to add texture, colour, form and structure to the urban landscape. Ornamental shrubs are used to define paths and borders, and outline garden areas in both public parks and residential settings. Screen hedges are used as living walls to provide privacy on residential property and are often manicured into decorative shapes in public parks and gardens.

The physical appearance of ornamental shrubbery depends on the particular species, the use made of it in the landscape setting, and the amount of maintenance it receives. The general growth forms include tall shrubs, woody ground cover and woody climbing vines.

Most ornamental shrubs can be found in public parks, gardens, private residential property and, to an increasing extent, indoors in various types of atriums located in the public areas of larger buildings and office towers. Ornamental plants can generally be categorized into three distinct types: coniferous needle leaf shrubs, broad leaf evergreen shrubs, and broad leaf deciduous shrubs.

Ornamental shrubs thrive in a variety of soil and moisture conditions. Generally, any area with moderate shade and slightly acidic, well-drained soils will suffice for the majority of ornamentals. Some species, such as the common or dwarf junipers, prefer well-drained soils low in calcium and high in magnesium. Others, such as Labrador tea, bog laurels and evergreen huckleberry (*Vaccinium ovatum*) favour cool, moist, peaty, and slightly acidic soils. A few such as the prickly pear cactus (*Opuntia sp.*) and big sagebrush (*Artemesia tridentata*) do better in drier, sunnier areas and are very drought tolerant.

Most ornamental shrubs occur in well maintained public and private settings, and the resulting intensity of management often precludes the coexistence of many wildlife species, which would normally utilize the wild growth forms for cover, nesting sites, food, and forage. However, some of the dense growth forms, such as hedges and sculptured bushes, may accommodate such nesting birds as Robins and House Finches. Many flowering species can also provide copious amounts of food in the form of nectar and berries for a variety of birds and butterfly species.

Ornamental shrubs rarely support large wildlife species, and in general, the less well managed and the more isolated ornamentals provide the highest wildlife values.

Landscaped Yards

Landscaping is the formal or informal planting of vegetation surrounding residences. The design varies with each site, but the overall affect is one of aesthetic appeal. Native or exotic species of plants are incorporated into fencing, rockery, and pavement layout in numerous ways, depending on the flair of their designer and the intended function.

Landscape design uses either native or non-native plants or a combination of both. Owners of residential properties typically combine native and non-native species for three reasons. Climatic conditions may not be suitable for non-natives and they often require more care and special cultivation techniques. For this reason, native species are preferred by some landscapers. Non-native species add a distinctive feature to gardens, often inviting different species of fauna, which are attracted to their unique colours, scents and nectars.

Different climatic, soil, and water conditions ultimately determine the quality of the species used in landscaping. However, plant traits such as size, form, flowers, foilage, and ease of propagation may be initial reasons for choosing a particular species.

Ornamental gardens typically display a large variety of exotics, and these plants are the feature attraction. As mentioned above, unique traits of exotic species invite many visitors, both human and animal. For example, the Botanical Garden at the University of British Columbia features plants from our own province and others from around the world.

Soil requirements vary with the species planted, and landscape designers will often provide the proper subsoil and topsoil prior to planting. Also, landscaping usually occurs on soils, which have been severely disturbed. It is necessary to modify the site by adding nutrients or new mixtures of soils in order to provide proper water drainage and aeration. Plants with deep root systems may be jeopardized if underlying subsoil contains rock or construction debris and they are unable to penetrate such compaction.

Floral variety will vary based on the individual taste of the landscaper and property owner. Combinations of annuals and perennials are common in residential and ornamental gardens. Perennials are considered mainstays of gardens (Musteg 1980). Daffodil (*Narcissus sp.*), iris, lupine, and crocus are examples of perennial flowers that are commonly used in landscape design because they are easy to grow and do not need to be replanted each year.

Annuals are also a favourite in all types of landscaping. They include plants with a maximum lifespan of one year and are usually exotic species that have been bred for their colour, foliage texture, full flowers, and ease of propagation. Ornamental gardens typically replant annuals as the seasons change because some varieties of these plants have life spans of only a few weeks and then wither and die. Examples include tulips, petunias, and marigolds (*Calendula sp.*).

By introducing a variety of native and exotic species into a garden, there will inevitably be a diversity of fauna. Trees, hedges, and shrubs provide cover, perching and nesting sites for birds and squirrels. They also serve as a food source by way of fruits, berries, nuts, and seeds. Different species of insects are attracted to the colours, scents and nectars of flowering plants and herbs. Gardens in particular entice a variety of species of butterflies and dragonflies. Attracting bees is also important in that their pollination is essential to a number of plants including apple and cherry trees.

Rooftops and Rooftop Gardens

When people think of a rooftop, they often envision a flat, hard surface, perhaps mixed gravel with tar and where access is often restricted for safety reasons. Rooftops offer tremendous views and vistas overlooking busy and bustling streets and the neighbourhood as a whole. Wind channels through corridors of buildings swish up to the rooftops. Those urban residents who are fortunate enough to have access enjoy the many pleasures of being high above the city.

Rooftop gardens can provide many benefits. Rooftop vegetation:

- Reduces urban heat island
- Removes air pollutants
- Makes building more energy efficient
- Improves aesthetics
- Increases roof life
- Reduces photochemical smog
- Reduces ground level ozone
- Provides urban recreation
- Absorbs sound
- Increases biodiversity

The National Research Council's (NRC) rooftop garden in Ottawa reduced heat entry into the building in summer by 85%. A rooftop of vegetation reduces runoff by 90% compared to other roofs. In terms of environmental certification for a building, a rooftop garden can earn one LEED point, with an additional point for Stormwater Management. Concerns about rooftop gardens include the loading on the roof from its weight (the solution to this is to use lighter soils containing Styrofoam) and water damage if the water barrier leaks.

Ecologically, rooftops are harsh habitats for many plants and wildlife. The dry, exposed growing conditions make them most suitable for grasses — they are sometimes referred to as "meadows in the sky." With maintenance, however, the rooftop can be converted to a viable garden.

Rooftops without gardens are generally hot and dry. The asphalt shingles, tar, and gravel absorb heat from the sun and reflect it onto other surfaces. In dense urban areas, rooftop gardens can potentially play an important role in modifying the rooftop climate for vegetation so that it is similar to ground level. An experiment was conducted on a flat roof on an old industrial building in Toronto, Ontario to explore alternative uses of rooftops (Hough 1984). The experiment showed that plants readily adapt to the hostile climatic conditions associated with rooftops. Further, the vegetation growing on peat moss media had kept the ambient air temperature above freezing during the winter. Rooftop gardens are becoming increasingly popular on apartment, condominium and office towers. Not only are they aesthetically pleasing, but they provide cool ambient air for the residents.

Gardens usually consist of boxes and different types of containers. They are placed to maximize sunshine or shade, as is required by the species. A community garden initiated by McGill University in Montreal, Quebec on a community

Rooftop gardens go a long way in softening the harsh features of a building.

centre roof in downtown Montreal proved to be a productive and practical way to utilize open spaces (Hough 1984).

Rooftops of older buildings may seem unsightly at a first, but may offer some interesting habitat types. Patches of grasses may colonize an area where soil formation processes have been initiated by lichens and mosses, providing habitat for insects.

Rock Doves roost in large flocks on the ledges of rooftops and build their nests in nooks and crevices. Old rooftops may succeed into naturalized habitats. Flat rooftops on large buildings may be used as nest sites by gulls and nighthawks. The eaves of roofs and the undersides of corner roof tiles are common sites for the European Starling.

Wind dispersed plants such as grasses, mosses, and some flowering weeds are common on old rooftops. Garden plants in pots and boxes are maintained on some buildings.

Natural Services Provided to Garden Landscapes

Garden landscapes provide many important natural services to the urban landscape. These landscapes can help reduce heating and cooling costs associated with homes and other buildings. Gardens and trees planted near homes, as well as vines and rooftop gardens, shade and cool buildings during hot summer months, and help reduce heating costs during the winter. All the plants associated with garden landscapes also work to improve air quality, as discussed in previous chapters. The flowers support pollinators, such as bees and butterflies.

14

Public Landscapes

Public landscapes are areas open for the general public to enjoy. They include parks and other open spaces that offer recreational opportunities and are usually operated and maintained by municipalities and other government organizations.

Even small spaces such as vestpocket parks can be important to wildlife and people. Kaplan and Kaplan (1995) refer to "directed attention fatigue" as a common condition affecting many of us in cities. The hustle and bustle of traffic, advertising, and other distractions means we have to concentrate harder on what we need to pay attention to in our lives. This may particularly be the case in our jobs which may be far less interesting than competing interests. A remedy for directed attention fatigue is "being away" from the cause. This requires involving oneself in a cognitive content different from the usual. Nature provides this "being away" experience. Kaplan and Kaplan found that the restorative experience of nature does not necessarily require a pristine natural setting. The proximity to nature, either perceived or actual, is however, essential. A distance just three minutes away can be overwhelming. Street trees and small parks can provide this proximity.

Parks

City parks range in appearance from highly maintained and manicured to natural and undeveloped. Depending on size, location and infrastructure available, a park provides various activities for people. Large parks may contain racquet courts, sports fields, petting zoos, or swimming areas. Smaller parks may only provide a playground or picnic area.

In general, city parks are maintained to whatever degree keeps them aesthetically pleasing. Urban parks retain a particular style, "landscape gardening, in which an enlarged corrected, refined, and idealized portrayal of nature is presented at an aesthetic level" (Gilbert 1991).

A city may develop around an extensively wooded site, allowing the park to become a green oasis. A park may also be created on an abandoned industrial lot or in an area unsuitable for development due to soil instability or gradient.

Parkland may be divided into two categories depending on the maintenance activities that occur: high maintenance and low maintenance. High maintenance parks are typically large and provide a variety of recreation opportunities for the public. City parks that are continuously maintained and manicured typically display areas of mowed grass containing scattered trees and are supported by areas of woodland. Ornamental landscaping such as formal beds of roses, annuals, perennials, and rockeries tend to occupy small regions, as they are expensive to maintain.

Low maintenance parks emphasize nature in an undisturbed manner and usually only have footpaths and trail systems. Trails are built throughout parks to provide public access.

Cycling, boating and enjoying nature are some forms of recreation in a park.

Soil compaction is typical in high maintenance parks along trails. Compaction may occur from human and vehicle traffic or through the addition of soil layers for landscaping. Compaction results in restricted soil aeration and increased soil strength, which may slow down root growth and affect the vegetation.

Wildlife diversity also depends on the degree of park maintenance. Low maintenance parks are able to support greater species diversity. Habitat value to wildlife also depends on size, structure, and availability of shelter. Bird abundance and diversity is especially dependent on shelter as it provides cover, nesting and perching sites. If not sufficient in number, these factors may limit the population size of certain species and allow for others to dominate the existing niches.

Wildlife diversity is greatly influenced by the type of plant community it supports. Vegetation provides seeds, roots, fruits, and insects. Of course, humans also play a role in contributing to the food chain in parks. By hand feeding wildlife or leaving garbage, animals such as squirrels and raccoons learn to supplement their regular diets of berries, seeds, and nuts.

Vestpocket Park

Vestpocket or mini parks are unique components of the urban landscape. They are small gardens at intersections or next to buildings. They can be found within secluded areas of apartment complexes or surrounding high-rise apartment buildings. They may also be found along side walkways, busy roads, and even in playgrounds for children that are available for public access.

The appearance of the vestpocket park depends greatly on the variety of plant species present. Selection is based on plant characteristics such as colours, textures, growth forms, whether or not they flower, if they are deciduous or evergreen, as well as their maintenance requirements. These factors determine the quantity and placement of each type of plant.

People value gardens and public green spaces in the city. To make living space more desirable in the city, architectural design has changed over the past years to incorporate aesthetics to implement visual health, vitality, and beauty in the urban environment. Pockets of vegetation, which are incorporated in new construction, compensate for the sterility of cement and steel, which stifles a city's character; in this way, gardens also generate a sense of life.

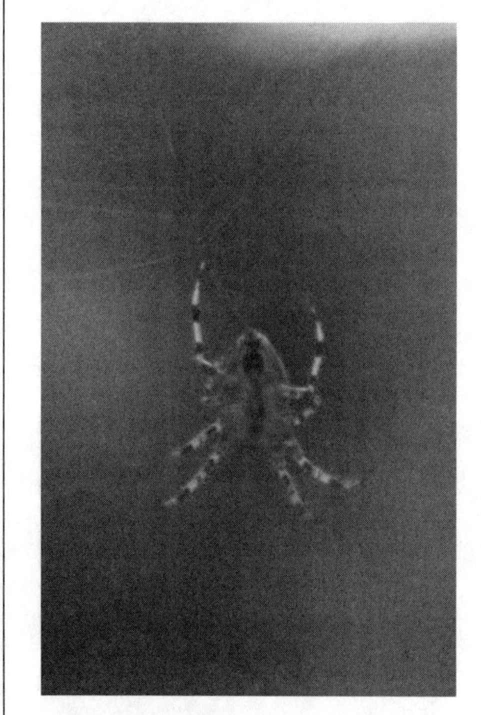

When looking for nature in a vestpocket park you have to think small.

Within vestpocket parks trees and shrubs are arranged in attractive formations to add beauty and greenery to the urban landscape. However, pockets of vegetation are not only important for their aesthetic appeal, they also establish the general character and framework of the landscape. More than any other plant, trees add personality to the landscape. They provide shade, give a feeling of shelter and form the focal point within the garden. Trees can also be situated to frame special vistas, as well as block less attractive ones (Sunset Books 1993).

Shrubs and ground cover form the base of the garden setting. They create a smooth transition between the trees' height and the ground level. They define boundaries, make trails, and create a lush appearance that influences the view. There are countless varieties of tree and shrub species, which offer seasonal benefits such as showy flowers, fruits and autumn foliage colour.

Vestpocket parks are not solely products of visual expression. These parks can be designed with nature in mind and contain more species diversity. They may attract wildlife into the city from larger surrounding parks. These pocket parks may serve as shelter and as passageways for raccoons, skunks and birds to movc about within the city. Public mini parks are not manicured frequently, so insect populations are able to flourish naturally, which attracts more birds. The flora and fauna in these important green spaces may be a person's sole experience of nature as they watch birds chirping in the trees or as they admire the clever designs of attractive foliage. Their experiences increase the value of the pocket parks immensely.

Golf Courses

Golf courses are large expanses of highly maintained lawns creating greens, fairways and roughs. They are beautifully designed and the layout of the course varies depending on the flair of the designer and the desired level of challenge to the golfers. Fairways are usually lined with trees and tall shrubs to separate them from other fairways. Putting greens, located at the end of the fairway, are very well groomed to maintain a smooth surface. Older, well-established courses are often popular mainly because of the attractiveness of the landscape. Golf courses can occur within city limits but are more common along the periphery, often on agricultural lands.

Golf courses offer some of the more valuable wildlife habitat on private property.

In addition to separating fairways, trees and shrubs also add a sense of peacefulness and relaxation appreciated by golfers. Some golf courses may also include ravines with streams. Vegetation and landscape features provide valuable wildlife habitat.

Golf courses invariably displace wildlife by removing the viable habitat and range area of large mammals such as black bears, cougars (*Felix concolor*), and deer. The remaining populations will wander back onto the golf course, requiring golf courses to implement large mammal management plans to minimize their encounters with golfers. For black bears, this plan consists of regular monitoring, educating staff on bear habits, avoiding the planting of their food types, regular removal of refuse, establishing procedures for dealing with problem wildlife and posting signs warning of wildlife. Coyotes easily adapt to changes in their range areas. They most commonly return to the adjacent forests and exploit the periphery of the golf course and fringes of residential areas. Deer remain residents of the forested areas along golf courses and may wander on to the golf course to feed on the young leaves of trees and shrubs. Grass turf of golf courses is favoured by field mice and voles. The roughs adjoining fairways and the forest edge provide an ideal habitat for mice and voles.

The edge habitat along the forest is often excellent bird habitat. Exotic vegetation such as fruit trees and colourful flowers planted throughout the golf course are also attractive to birds. Fruits and berries provide food for many songbirds. These songbirds also roost and nest in the trees and shrubs surrounding the fairways and throughout the course. Installation of nest boxes further enhances the golf course as songbird habitat.

Ponds and wetlands are easily constructed in golf course design. These habitats attract various waterfowl and marsh bird species, such as Mallard, Wood Duck, Pintail Duck, Great Blue Heron, Red-winged Blackbird, and Marsh Wren (*Cistothorus palustris*).

The types of plants found in a golf course depend on the flair and knowledge of the designing architect. A golf course can be planted with a mixture of native and non-native vegetation or strictly native. The current trend in golf course design is to incorporate existing landscape features and plants into the design. Audubon International offers certification for golf courses through its Audubon Cooperative Sanctuary Program for golf (www.audubonintl.org). It provides information that will assist golf courses with:

- Environmental planning
- Wildlife and habitat management
- Chemical use reduction and safety
- Water conservation
- Water quality management
- Outreach and education.

Cemeteries offer green spaces that will be preserved in perpetuity. Although some cemeteries are municipal property, most are private property. Nevertheless, they are considered a community asset for the green space they provide.

Cemeteries

Cemeteries provide lawn habitat and may have many of the features found in golf courses. Most have trees and shrubs, especially along the perimeter of the property, and some even have ponds and other water features. Vegetation is usually planted for aesthetics and as a screen to separate adjacent land uses. The landscape in cemeteries is often highly groomed. Older cemeteries have mature trees.

Post Secondary Institutions

The vegetation of a university or college campus is mainly grass, with some trees and shrubs. There may also be some ponds or streams. This habitat can provide resting and feeding stops for wildlife traveling between parks or other larger green spaces.

Airports

Airports have varying levels of flight activity. They are often situated away from the city in undeveloped areas with a high wildlife potential. Wetlands in particular are often found near airports, because they are generally left undeveloped, and the aircraft would not disturb people.

Post secondary institutions not only offer green space in their playing fields and landscaping; also, they often have undeveloped sites endowed to them that, although intended to provide space for future development, may provide wilderness for biodiversity for many years.

The habitat consists mainly of level land of grasses or old field habitat. The grass areas are important for small mammals such as mice and voles. These in turn are fed upon by hawks and owls.

It is important to note that at airports the primary concern is safety for the aircraft and wildlife is not encouraged. Habitat management plans usually address methods to discourage wildlife. For example, at YVR Vancouver, British Columbia, habitat management by the Airport Authority (Vancouver International Airport Authority community update newsletter November 2002) includes:

- Mowing grass to reduce waterfowl grazing
- Planting grass (reed canary grass) and non-berry producing shrubs and trees that offer little food for wildlife

- Draining sites and grading land to eliminate pools of water that attract waterfowl, gulls and shorebirds
- Eliminating perches and roost sites
- Inspecting buildings to identify and eliminate nest and roost sites
- Conducting patrols and use of pyrotechnics, sirens, lights, propane cannons and other devices to scare away wildlife

The large open areas associated with airport runways may provide extensive grassland habitat.

However, airports often own property not directly used in aviation. It is these areas that are the most valuable to wildlife (Aguilar 2003).

Community Gardens

Community gardens are important and beneficial initiatives in neighbourhood landscapes. The areas covered by community gardens are usually small, often just one or two hectares. The land is subdivided into plots and is available to community members who agree to participate in the management of the site. Committees of gardeners usually do the planning and make the decisions about the gardens.

Community gardens are an economical alternative for obtaining fresh, organically grown vegetables. They were originally designed for people living in apartment buildings and who did not have access to land, so that they could garden. Over the years, residents of urban centres have been drawn to the community gardens because of the sense of community they foster.

Each plot may vary, simply because of the various choices different gardeners make. Essentially, the soils must be productive and fertile, and the vegetables and flowers must be watered and weeded regularly.

There is a great diversity of plant species within a community garden, although largely not native. This diversity provides habitat for a wide range of butterflies and birds.

Success in gardening stirs a sense of pride, joy and achievement. Children often find gardening fun and an interesting way to learn about nature. Community gardens are especially important in inner city circumstances where the land base available for gardening is scarce. The community gardens are a place where people can get acquainted and share and compare ideas. It fosters a feeling of comradery and neighbourliness.

This community garden is frequented by many hawks and owls.

Nature's Services Provided by Public Landscapes

Public landscapes collectively cover large areas and a wide range of habitat types. This size and diversity makes them important in providing all of nature's services mentioned in previous chapters — generation and renewal of soil and soil fertility, cycling of nutrients, pollination of crops and natural vegetation, biological control of agricultural pests, dispersal of seeds and translocation of nutrients, fisheries, maintenance of biodiversity in support of agriculture, protection from the sun's harmful UV rays, climate regulation, moderation of temperature extremes and the forces of wind and water, and reducing pollution.

However, their public access makes them even more important for the services nature provides in maintaining our emotional and psychological well being. They provide aesthetic beauty and intellectual stimulation that lift the human spirit.

Part 3

Improving Biodiversity

Suggestions for Backyards and Balconies
Government Action
Case Studies
A New Urban Reality

15

Suggestions for Increasing Biodiversity in Your Yard or Balcony

Small plantings in yards and balconies are invaluable when removing pollution from the atmosphere. These plantings also provide important wildlife habitat, which is becoming smaller and increasingly more fragmented in cities.

Much can be done to attract wildlife to yards and balconies. Planting certain types of flowers, shrubs, and trees in gardens provides essential life elements for a variety of wildlife. By attracting wildlife to your yard, you will begin to feel a connection with nature and be provided with hours of enjoyment (Arbuckle and Crocker 1991).

Even though your plantings may not seem like they go very far to make a difference, the cumulative effect of many people planting their yards and balconies significantly improves air quality and wildlife habitat.

This chapter will help you to create wildlife habitat in the space you have available. We will begin by discussing wildlife's requirements for life and the major principles of a native plant garden. We will then show you how to create your own garden. Finally, we will give you suggestions on how to attract certain wildlife species to your yard and theme gardens. Plant suggestions in this chapter are native to the Lower Mainland of British Columbia. Plant suggestions for other regions can be found by contacting local natural history organizations or the Ministry of Environment.

Four Elements of Habitat

Wildlife requires four basic elements: space, food, water, and shelter. These four elements are collectively known as wildlife habitat. By supplying these elements, you are on the way to attracting wildlife to your yard.

Space

All wildlife requires three-dimensional space as part of their habitat (Pearman and Pike 2000). Spatial requirements vary greatly, depending on the species:

some birds, such as tree swallows, can live in one residential yard whereas a pair of northern flickers require a much larger home range (Pearman and Pike 2000). Even if you have a small yard or balcony, it still functions as a very important part of many species space needs. Of course, the larger your yard, the more species you can expect it to support. Your backyard habitat connects itself with other yards and balconies in your neighborhood to provide important corridors for wildlife to move throughout different habitat.

Food

Plants that produce berries, seeds, nuts, and nectar are the main food source for wildlife (Schneck 1993). Food plants are the foundation of the food chain (Pearman and Pike 2000). A leaf may provide food for an insect, which then becomes food for a bird, which then becomes food for a larger bird or mammal. It is also important to plant both evergreen and deciduous species, as well as those that bloom and fruit at different times of the year. Planting a diversity of vegetation helps support a wide variety of wildlife. Feeders are a great way to supplement natural food sources (Adams 1998). They are more important in the late fall, winter, and early spring when there are not as many natural food sources available. Since different species eat different foods, it is a good idea

Look Who's Knocking

Sometime when you're out back in your yard you might hear some knocking in the trees. It might be a loud rat-a-tat, or a solitary crisp rap or quieter, gentler tap. The knocking could be one of the several types of woodpeckers that share our residential neighbourhoods. The most spectacular one is found more in the suburbs or by ravines. It is the Pileated Woodpecker. As large as a crow, it is black, sporting a large red crest and wide white wing patches when it flies. When looking for food it drills large, oval-shaped holes on tall trees, some of which it will excavate deeper for sleeping and nesting.

You may have noticed a large bird on your lawn, feeding with the robins. Much larger than the robin and with shorter legs, you might think, what's that? Well, it's an unusual place for a woodpecker, but it is probably a Northern Flicker. This pigeon-sized bird is gray-brown with dark spots and reddish shafts on its wing feathers. Much more common than the Pileated, it has a similar loud, piercing call that it cries out while flying off in the wide w-shaped arches typical of flight for all our woodpeckers.

You may have noticed a row of small holes in straight lines on the trunk of a tree that look like they were made by a machine gun. This is the work of the Red-breasted Sapsucker. It will come back to these holes to check for bugs that were attracted to the sap in the holes.

Our smallest sparrow-sized woodpecker is the Downy. It moves silently through our trees and shrubs, inspecting them for pests that make a tasty meal. Males have a bright red patch on the back of the head. The Downy Woodpecker frequently accompanies flocks of chickadees, kinglets and nuthatches that tend to move together as a group for some extra protection in noticing predators. Another woodpecker that looks much the same, except it is a bit bigger, is the Hairy Woodpecker. Both the Hairy Woodpecker and the Pileated are indicators of good ecosystem health, and that means a healthy place for us to live, too.

You can attract woodpeckers to your yard with suet, either stuffed into a crack or hole in a tree or in a wire cage. Suet is available from most pet stores and large home hardware centres.

to experiment with the seed mixtures used in your feeder (Campbell and Pincott 1995).

Water

All organisms require clean and fresh water for drinking and bathing (Campbell and Pincott 1995). Many amphibians and insects also complete part of their life cycle in water (Campbell and Pincott 1995). It is important to ensure that your water source is close to shelter so that wildlife using it feels protected. Water can be provided in yards as ponds, streams, birdbaths, and puddles.

Shelter

Shelter is an important component of wildlife habitat. It helps protect many creatures from predators, wind, and rain (Adams 1998). Shelter also provides places to nest and roost (Arbuckle and Crocker 1991). Most species require a variety of shelter when nesting and feeding. Plants are the primary supplier of shelter (Schneck 1993). Through urban development, many nesting sites are dis-

The Night Shift

Many of us notice the animals in our yard during the day. We enjoy watching the songbirds and squirrels snooping around, looking for treats in our gardens or on our balconies. We also know that bats fly around at night. But who is there? How do we know if there is anything around?

One of the more obvious signs of the night shift is a lawn that has been shredded, with numerous little clumps of overturned grass. This is the work of skunks that are looking for food amongst the roots. What they have usually smelled that made them dig was an inviting buffet of abundant larvae of teh European chater (*Rhizotrogus majalis*), a brown beetle resembling a small june Beetle..

Another night visitor that is rarely seen is the northern flying squirrel. If you have a red Douglas squirrel in your neighbourhood, chances are that you have a flying squirrel, too. They often occur together in equal numbers in the same areas, the Douglas squirrel working by day eating seeds from cones and buds, the northern flying squirrel working by night, hoping to find some truffles hidden in the ground.

Ravines in your neighbourhood are the daytime resting places for owls. Barred and Saw-whet Owls in particular are relatively tolerant of people and are not uncommon visitors in residential areas at night. Sometimes they will linger in your yard in the morning to roost in a tall tree or shrub, but the crows will soon find them, mobbing them until they leave for a more secluded area.

Let us not forget moths, perhaps the easiest of the animals of the night shift to see. They are so diverse and remarkable to watch that it doesn't matter that we don't know their names. The small, T-shaped comma moth clings to the wall next to a light on your house. The medium-sized sphinx moth is shaped like a stealth bomber and commonly flies into our house through an open door or window. The larger brown polyphemous moth clings to light fixtures. It is the size of the palm of your hand with large wingspots that are supposed to fool us into thinking they are the eyes of an owl.

Then, of course, there are the raccoons and coyotes. They tend to be the shiftiest of them all.

appearing. You can provide artificial shelter in your yard for birds, bats, bees, and butterflies.

Each species has a different definition of optimum habitat and requires different combinations of the four elements (Schneck 1993). The diversity of species you attract to your yard depends on the amount of space you have as well as the variety of food, water, and shelter.

Native vs. Non-Native Plants

Native plants are those that have existed in a particular region prior to European contact (Pettinger 1996). Plants that are native to an area have adapted to the climate of the region and have co-adapted as food sources for native birds and wildlife. Native plants provide the best overall food source for wildlife and are recognized habitats (Carr 1999). These species require less maintenanceltd — fertilizer, water, and pest control — than non-native or ornamental plants. Native plants also support more wildlife species than non-native plants (Levy et al. 1999).

Periwinkle (*Vinca major*) is an invasive alien species that competes with native species such as violets.

Many people feel that native plant gardens tend to be unattractive and colourless. Many native plants bloom in spring. However with some research you can discover a wide range of native plants that ensure your garden is colourful throughout the growing season. The flowers of many native wildflowers and shrubs are very attractive and provide colour throughout the year. For a variety of colour and species, you can also combine native and non-native species for a mixed garden (Pettinger, 1996).

Plants termed as non-native, exotic, or introduced are those that are grown outside their native range (Pettinger 1996). Non-native species tend to have less food value than native species, as many species do not produce berries, seeds, or nuts (Levy et al. 1999). They compete with native species for growing space, water and nutrients, resulting in poorer growing conditions for native species.

Some non-native plants are highly invasive under the right conditions. Their natural predators are not present and many of these species are hardy and aggressive (Levy et al. 1999). In southwestern British Columbia, purple loosestrife, ivy, and Scotch broom are three examples of species that have been introduced to the region and have invaded some areas. Invasive species reduce biodiversity by outcompeting a variety of native species. Reed canary grass (*Phalaris arundinacea*) is an aggressive invasive in perennial wetlands.

How to Obtain Native Plant Stock

Native plants can be acquired through nurseries or can be propagated with seeds and cuttings. You might also be able to salvage some plants from an area about to be bulldozed for development. In the past native plants have been difficult to purchase through nurseries; however, many plant nurseries are now stocking increasing numbers of native species due to increased demand. This makes obtaining native plant stock easier. When purchasing native plants from a nursery, it is important to ensure the nursery is reputable and does not get its plant stock from the wild.

Propagating plants from seed is a great way to obtain native plants and reduce costs. If you are collecting seeds or cuttings from the wild to propagate ensure that you only take 10% of seeds from the plant (Levy et al. 1999). Leave the rest of the seeds and fruit as food for birds and for the plant to disperse itself. Make sure seeds from your immediate region are used (Levy et al. 1999), as they are best adapted to the local environment.

In British Columbia, three species are protected by law and it is illegal to take any part of these plants: western flowering dogwood (*Cornus nuttallii*), Pacific rhododerdron (*Rhododerdron macrophyllum*), and western trillium (*Trillium ovatum*), (Pettinger, 1996). It is also illegal to remove any part of a plant from any park or protected area.

Native plants being propagated as a high school science project. Red-osier dogwood is easily grown from cuttings.

Creating A Backyard Wildlife Habitat

Creating a wildlife garden in your yard can be very rewarding. As your garden grows and matures, a diversity of species will be attracted to your yard. Remember, you will only be able to attract wildlife that naturally occurs in your region and species that are well adapted to urbanization and are tolerant to disturbance. Unless you live near a forest edge, it is unlikely that large mammals, such as deer, will be attracted to your yard. When creating backyard habitat, it is important to assess what already grows in your yard, form a plan, and carefully select plants.

Attracting wildlife to your yard can lead to many new discoveries, like this snake.

Outline Your Yard

Before getting started with your garden, it is important to assess what you already have in your yard. A good place to start is by creating a base map. A base map allows the existing conditions of your yard to be evaluated. Important components of a base map are the location of buildings, patios, driveways, utility lines, existing vegetation, water sources, existing feeders and nest boxes, brush piles, rock shelters, stumps, logs, snags (dead trees), wildlife travel corridors, and wildlife viewing areas (Link 1999). Once your base map is complete, you can determine how much space you have to work with.

Evaluate Existing Conditions

- Your site's environmental characteristics need to be evaluated prior to beginning designing your garden. (Pettinger 1996). Some important questions to ask yourself when planning your garden are:
- Where are your "people areas" (i.e., children's play area, deck, patio, storage shed)? The movement and activity of humans disrupt wildlife habitat — it is best to plan habitats away from these areas.

Even a habitat such as a rotting tree stump can support many species of fungi, mosses and invertebrates.

- Do some of the plants in your yard already provide food for wildlife? Note any shrubs and trees that have berries, cones, or seeds. Also note any flowering plants and when they bloom.
- How many of your plants are native?
- Count the number of different kinds of plants in your yard. Is there a high diversity (a high number of different types of plants)?
- How are plants located? Are they isolated from each other (planted in formal rows or single tree)?
- Is there a mix of deciduous and evergreen plants?

Backyard Politics — Hawks and Doves

Enjoy a good political debate? Spend some time outside in your neighbourhood and watch the actions of hawks and doves govern the animal world. The Cooper's Hawk likes "take out" food that it can pick up on the go. It especially likes House Finches and House Sparrows that it can snatch in mid-air and take to a tall tree to eat while it does some people-watching from its secluded perch. If you've seen a crow-sized gray hawk perched in your yard, chances are it's a Cooper's Hawk. Sometimes it is so cheeky it will sit next to your bird feeder, impatient for its own next meal to come by.

Another city hawk is the Red-tailed. It tends to perch on tall trees and lamp posts. It's "kyerr" cry is the one frequently heard in the city (and in movies). It especially enjoys the view from lamp posts along roadsides, watching us stuck in rush hour traffic, "sitting ducks", so to speak, thinking that we would make a great dinner if only we weren't so big and able to defend ourselves.

At the other end of the birds' political spectrum we have the doves. Especially in the suburbs we have Mourning Doves visiting our yards and feeders. Mourning Doves have a soothing "cooing" call that is particularly noticeable in the morning or in the middle of a lazy sunny summer day. Mourning Doves have a large, white-fringed tail that, in flight, looks like a prehistoric archaeopteryx.

The native Band-tailed Pigeon prefers to fly in pairs. It is actually more of a woodland bird. It is mainly slate gray in colour, with two dark bands on its tail.

Then there is the common pigeon, or Rock Dove. In its natural habitat in Europe it nested on cliffs. In the city, our tall buildings have become its cliffs and spilled grain from railcars or bird feeders are its new diet. Normally gray with darker markings and an iridescent head, domesticated forms have been bred to produce as many varieties as dogs with plumage including mottled brown and tufted. Two of the varieties are the beautiful white "peace doves" that circle overhead at event ceremonies such as the opening of summer Olympic games, and the homing pigeon that delivers messages for us.

- Are there any unused lawn areas?
- Where do you find wildlife in your yard now? At what times of the year?
- Where are the sun and shade areas?
- Is the soil naturally dry or moist?
- Is the soil fertile? Acidic?
- Where are the desirable and undesirable views? Check this from many different angles.
- Where are the wet and dry areas of your yard?

Form a Plan

Once your yard has been evaluated, you can begin creating your garden plan. When beginning to plan your yard it is important to think about what type of wildlife you wish to attract. Ideas on how to plan for specific wildlife are provided later in this chapter. Plants need to be chosen carefully to meet the conditions of your yard as closely as possible. Keep in mind views, people areas, and existing vegetation when forming your plan. Human activities should be concentrated so wildlife can inhabit their areas of your yard without being disturbed by constant traffic. Existing vegetation, especially trees, should be left in your yard, as older vegetation supports different species than young vegetation does. Other factors that should be considered when designing your plan include diversity of plants, planting in groups, layering, and edge effects.

Diversity

As the complexity and diversity of your yard increase more wildlife will be attracted. A variety of plant species allows greater choice of food and shelter.

Planting in Groups

In the wild plants are often found growing in clumps or groups. Designing your garden to have clumps of plants instead of rows or individual plants is beneficial to wildlife, since clumps are a concentrated source of food (Pearman and Pike 2000). Animals will usually stay longer at a clump of plants that provide them with food than at individual plants, since there is more to eat.

Layering

Different wildlife require different layers or levels of habitat. Examples of layers are groundcover, perennials, low growing shrubs, tall shrubs, and trees. Some species live underground, others on the ground or in low shrubs, and others live in treetops. By providing different layers of vegetation, you can meet the needs of many species. When planning layers in your garden, make sure the shortest plants are in the front, near viewing areas, and the tallest plants are farthest away (Pearman and Pike 2000).

East Meets West Out Back

Did you move out west from Quebec, Ontario, New York? Do some of the animals in your yard look familiar? Well, they are. They got here from eastern Canada and the United States before you; or perhaps you are mistaken and they are not what you think.

The eastern gray squirrel has prospered in urban areas. It likes the deciduous trees and other habitat features of residential areas. As the cities grew and housing spread, so did the gray squirrel, benefiting from the ill fortunes of native western Douglas squirrel that needed the original coniferous evergreen trees. The gray squirrel also has a black phase that is quite common. Cute as they are, try to go easy on feeding the gray squirrels — it will cost more than a few peanuts to get rid of them if they make a home in your attic, or the attics of your neighbours.

The western spotted skunk does not like cities, but the striped skunk that is found over most of North America likes them just fine. It is the striped skunk that we see (or more often smell) in lanes and yards. Once thought to be a member of the weasel family but now considered to be in a separate family (Mephitidae), the striped skunk likes to live under porches or garages and has a diet that covers everything from fruits and vegetables to meat. Some striped skunks are real city dwellers, preferring the fast pace and night life of a city's downtown.

Then, there is the case of mistaken identity. The western loud Steller's Jay is the provincial bird of British Columbia. It is dark blue on the lower part of its body and a dark brown or black on the upper part and head. It has a large peaked punk-like crest (no hair gel required). People who recently moved to the west from the east often mistake it for a Blue Jay, which is more familiar to them, but is actually a different bird. The Blue Jay is light blue in colour with white spots and a crest that is not so big. The Blue Jay is considered to be a pest in many cities, eating the eggs and young of birds to such a degree that they are considered to be a greater threat to songbird populations in cities than cats. The Steller's Jay prefers hazelnuts to meat, in keeping with the more vegetarian and healthier western lifestyle.

Edge Effects

For wildlife gardening, an edge is where two different habitats meet: as when forest meets grassland. Edges are beneficial to wildlife on a small scale, such as in yards, since a high amount of wildlife activity takes place at edges. The amount of edge in your yard can be increased by curving plantings, alternating clumps of planting with open space, and by graduating the height of vegetation in your yard.

Select Plants

Plants should be chosen carefully to meet your yard's conditions. They should also be specific to the wildlife you wish to attract and the habitat types that already exist. A list of native plants and the wildlife that use them are listed later in this chapter. Plants should be grown in groups, with species requiring similar site conditions planted together. As discussed earlier in this chapter, native plants are best at attracting wildlife to your yard, as they support a greater diversity of wildlife.

Creating Wildlife Habitat on Balconies and Other Small Spaces

Challenges of Balcony Gardening

Although many balconies are limited in space, they seldom reach their wildlife habitat potential. A stroll through the downtown area of any city offers many "blank canvas" balconies just waiting to house a few native plants, a nesting box, and a small birdbath.

Many of the principles of creating backyard habitats apply to gardening in small spaces, but there are a few practical things to consider when creating a balcony habitat. Due to the size and engineering of many balconies, consideration must be given to the height of plants once they reach maturity as well as to the weight of the plants, soil, and containers. Ask your landlord or building manager if there are any weight restrictions on your balcony, and then overcompensate when calculating the combined weights. Plastic containers and peat-based soils will reduce the weight of your balcony gardens. Many of the Lower Mainland's native shrub species do not reach tall heights and, with periodic pruning, your balcony can be kept under control.

Another consideration to go along with the height and weight of your plantings is the ability of your plants to withstand drier conditions. Most balcony plantings will not receive rain and are restricted to the supply of water that can be stored in relatively small containers and may be exposed to drying winds. Balcony gardens will have to be watered more frequently than indoor plants.

Hanging Gardens of Babylon — The Wonders of Fast Food

Is your balcony looking a little bleak these days? Just have a chair out there or do you just use it for storage? Look again! Your balcony could become a garden that can be part of one of the seven wonders of the world. Imagine if everyone with balconies grew a few flowers and vines. Balconies altogether can represent an area equal in size to a large urban park. If they were all planted, we could create our own modern Hanging Gardens of Babylon.

There are many types of balcony gardens. Hanging baskets are ideal for creeping and hanging plants and easy to set up. Window boxes are perfect for any balcony garden, even if you don't have an actual balcony. Wall plantings of climbing plants take advantage of the vertical space. Raised garden beds are sometimes best for rooftops and are basically larger versions of a window box. Or, plant an alpine trough (you may be living on one of the higher floors in your building after all), using plants like dwarf shrubs, heather and mosses that normally are found at higher altitudes.

You might be surprised to find a robin or House Finch nesting on your balcony one day. Someone once had a House Finch nest in a hanging basket on the 26th floor! Perhaps not the brightest bird in the flock — flying up and down 26 floors every few minutes to feed young? Why? Anyway, it worked out in the end. Your plants could also provide a tasty meal for many birds passing through, especially during migration when they're in a hurry and need fast food.

What Container Should You Choose?

There are a few general rules to follow regarding your container choices, but generally it all comes down to a matter of taste. If your plantings are deciduous and leave your containers bare in the winter, make sure you like the look of the container on its own. In terms of the material of your containers, there are a few considerations to be made. Wood, such as cedar, is an excellent insulating material, and may keep your soil from freezing over the winter, but make sure that it has not been treated with any toxic preservatives, and comes from a sustainably managed forest. Plastic containers are light-weight and offer the advantages of retaining more water, and being durable and inexpensive. More attractive than shiny plastic, terra cotta and clay pots are very common. Clay pots breathe, which means compost will not overheat, but your plants will require more watering.

There is a wide assortment of container styles to use on your balcony. The standard choice is often a type of free standing pot that can create a very beautiful effect if different sizes and shapes are clustered in a corner. Troughs or window boxes holding several plants can be hung along the railing or placed along the edge of the balcony. When these are filled with a combination of bushy plants and hanging vines, they can create soft edges for your balcony, and create habitat without taking up a lot of space. Tower planting containers are increasing in popularity. These tall, thin containers have holes for plants around all sides of the container, as well as an open top for vegetation. These containers are fitting for the balcony because they do not take up a lot of floor space. You can also think of them as a smaller version of the big picture. Tower plantings look like tiny apartments where everyone has decided to create their own garden balcony.

Hanging baskets offer another option for the balcony gardener. The ideal placement is in a partly sunny location, which means avoiding north and south-facing walls. Try to place them so they are sheltered from strong winds and, when fastening them to the walls or roof, keep in mind that a well-watered hanging basket can weigh up to 25 pounds.

Dirt Discussion

For the soil of almost all your container plantings you should combine the following: two parts peat, four parts loam, one part sharp sand, and chalk (five grams per bucketful, not necessary for lime-hating plants).

Since you will be growing plants in a contained amount of soil you may need to include slow release fertilizers. Tablets, cones or sticks are available to insert in the compost, but we would suggest you replenish the compost when plants begin showing mild signs of nutrient starvation.

Planting and Potting Techniques

The first step in any balcony planting is to make sure that your container has the proper drainage holes. The bottom of the container should have a 2 cm diameter hole every 10 to 15 cm. The container should then be soaked overnight and set in its permanent location before filling the next day. Make sure the container is standing on blocks or short legs (bricks work very well), to raise the drainage holes above the surface (Hessayon, 1996).

Starting to fill the container requires the covering of the drainage holes. These holes should be covered with a fine mesh or, perhaps, a stone. This allows the loss of water, but not soil, through the holes. If the container is large, add a layer of peat to reduce the amount of compost needed. Ideally, the compost layer should be no more than 23 cm deep. When compost is added, press it firmly with your hands and add it in layers if the container is larger. For watering purposes, leave a standard (this is recommended for any container) one inch between the top of the container and the soil surface.

When the container is full, you may begin planting. Remember to start in the centre, usually with the largest plant, and work your way to the outer edges. Plants should be placed closer together than they would be in an outdoor garden. Take the largest plant and dig a hole in the soil large enough for the plant's pot to sit in. Remove the plant from the plastic pot by placing it on its side and pushing gently on the rim of the pot. Rotate and repeat this until the plant can be removed by carefully pulling the pot away, while holding on to the plant at the very base of the stem. Massage the roots to loosen the dirt and root ball. Place the roots in the hole so that the plant base is level with the surrounding soil. Fill in the space around the roots carefully and press the soil down firmly because air spaces can allow fungus to grow. Repeat these steps for each plant and water thoroughly when you are finished all the planting.

If roots begin growing through the drainage holes, your soil is drying out very quickly; or, if your plant is growing very slowly despite favourable conditions, it is probably time to repot. For trees, shrubs and perennials, it is usually necessary to repot every three years, increasing the pot diameter by 8 cm. Note: If you do not want the plant to grow any larger, do not repot it. Water the plant and wait one hour before gently removing the plant from the container. If you are experiencing difficulty, run a knife along the edge, between the soil and the pot. Fill a new, larger pot with compost, and plant as directed above. After you are finished, tap the pot several times on a hard surface and water thoroughly. If the weather is hot and dry, keep the plant in the shade for one week.

Wall Plantings

There are many holders that are designed to be attached directly to a flat surface. These containers are found in both open and closed versions, and can be planted using the same methods outlined above for hanging baskets. The advantage to using a wall mounted container over a hanging basket is that the planting can look attractive before it has reached maturity. It is very important that you attach these containers firmly, and take into account that soil will be heavier after watering.

When considering designs for bare walls, one of the most exciting ideas is creating a plant tapestry. These vertical gardens, built against a sturdy concrete wall, require strong support and anchors. A heavy plastic liner is fastened to the entire wall. Starting from the bottom, begin attaching a large sheet of chicken wire (5 cm grid) to the wall with a thin, wooden strip running the length of the wall. Line a lengthwise area of the chicken wire with moss, and then a plastic liner, and fill with soil. This should create a semi-circular bulge away from the wall when a second horizontal strip of wood is placed across the wall approximately 10–30 cm higher than the last strip. Continue this pattern to a desired height. Remember that the plants at the top will have to be watered. To plant the tapestry wall, poke holes through the wire screen, moss, and plastic liner to the soil. These walls have a disadvantage of needing growth time to look beautiful.

Window Box Plantings

It is often recommended that window boxes be chosen so as not to detract from the plants, and to be in keeping with the style of the house. Stained or painted wood is a good material, because it insulates well, and will not be too heavy to properly attach it to the wall. Use larger screws and brackets to increase the weight that can be held. Make sure the window box has adequate drainage by drilling 2 cm diameter drainage holes every 10 to 15 cm in the bottom of the container. A drip tray can be filled with gravel and fitted below the drainage holes so that water does not drip down the wall and stain it. If you are ambitious you could place a drip tray 25 cm below your window box. This would allow birds to use the excess water, but it will have to be cleaned out periodically.

Always attach the empty box to the wall before filling and planting. Cover the drainage holes with stones or a fine mesh screen. Add 3 cm of gravel to help drainage. If weight is a concern you can reduce or omit the gravel layer. Add moist soilless compost in layers, pressing each layer firmly with your hands. For planting your window box there are three methods that are used most commonly. Plants can be:

(a) **Directly planted.** Plants are removed from the pots or trays they were in and placed directly in the compost-filled box. With this method newly grown plants are on display while they are becoming established.

(b) **In-Pot planting.** The plants are not removed from their individual containers, but are placed in the window box, and surrounded by gravel or peat moss. This allows you to easily remove and add plants throughout the seasons.

(c) **Liner planting.** Some window boxes are fitted with liners. These liners are usually plastic containers designed to fit the window box. If you keep a spare liner this can be removed, filled with plants and rotated between growing and display plants. Window boxes will need plenty of water, so it is best to keep them accessible.

Window boxes can easily double as temporary bird habitat. Although it is unlikely you'll have a pair nesting, if you leave your plants to decompose some insects may take up residence, and birds will visit to feed on them. Window boxes can also become useful feeding stations. If you place seeds in amongst the plants and have filled the soil to an inch below the container lip, you will avoid the seed scattering that may be caused by messy eaters.

Rooftop Gardens

The challenges faced by rooftop gardeners can also affect people who have balconies above tree level. Many birds perch in trees or on light posts and, if you are above these things, they may have trouble finding or getting to your habitat offerings. Recently, there have been some concerns regarding roof materials and birds that often nest on rooftops. Asphalt shingling and other dark, heat trapping materials can absorb too much solar heat and create problems for all life, including plants.

The principal components of a rooftop garden that covers the entire roof begin with the necessary structural support to hold the extra weight. A vapour controlling barrier is first placed on top of the roof, covered by thermal insulation, a support panel and a waterproof repellent membrane. A drainage layer is put down next, covered with a filter membrane. Only then is the roof ready for the growing medium and vegetation.

Creating an Alpine Trough

Taking your gardening to new heights on the rooftop may inspire you to plant alpine or rock garden plants. Alpine gardens can be housed in reconstituted stone troughs or an old glazed sink (which could be covered with a bonding agent and a mixture of sand, cement and peat). Make sure that your container has adequate drainage, and raise it by placing it on bricks in a spot that will receive a fair amount of sun. If you are using a sink, slant the container so that the water will run out the drain.

The ideal time to fill an alpine bed is in the late spring. The plants will need time to establish themselves before bad weather, so try not to plant in the late autumn through the winter. Cover the drainage holes with a stone or screening to keep the soil in the pot. Then test to ensure that water can still drain freely. Add a layer of gravel/rubble to a depth of 5 cm.

Add the planting mixture to within 2.5 cm of the top of the container, wait for 24 hours to see where it settles, and top off if necessary. A recommended alpine planting mixture consists of blending one part topsoil, one part peat (or rotted leaf mold), and one part grit or stone chipping. If the species you have chosen do not like lime, then use a mixture of one part topsoil, two parts sphagnum peat, and one part lime-free gravel.

Landscape the surface with two to five stones. Try to find rocks that have deep cracks and are already covered in lichen. Bury each stone to one third or one half of its depth, and wait two weeks to begin planting. Conifers do very well in alpine troughs and may be a good starting point for your planting. Be careful to leave some room for spreading and try to combine mat-forming plants with perennials and trailers for the edges. Water after the planting is complete, and cover any bare soil with stone chipping. Continue to water your alpine garden when the top inch or two of the planting is dry, and water until you hear it draining out the bottom.

Raised Garden Beds

Similar to the alpine trough, raised beds tend to be larger. They offer the same advantages of adding a dimension to a flat rooftop, and are easy to attend to. When considering your dimensions, make sure you can reach the middle for watering and weeding from the outer edges. For construction material, 2 × 4 wood is ideal, as are old railway ties or bricks. With any of these materials you can leave gaps in the side walls and fill these with plants, creating container walls with interspersed vegetation.

Once the walls of the raised bed are constructed the bottom half should be filled with a layer of rubble, stones, or broken bricks. Fill the remainder of the bed within an inch from the top.

Maintaining Your Wildlife Garden

Although native plants are adapted to your region's environment, they still need to be cared for and maintained. If left unkempt, invasive species can take over your wildlife garden.

Bark Mulch

Using bark mulch around your plant will decrease weed growth and increase the presence of beneficial insects. Bark mulch holds moisture in the soil, reducing the amount of water you need to use in your garden. It also insulates plants during the winter, protecting them from frost. As bark mulch decays, nutrients are added to the soil and plants are fertilized.

Weed Control

As mentioned above, using bark mulch will decrease the likelihood of weak growth. Planting native ground cover also helps to control weeds, since they can take over ground space before weeds can grow. To remove weeds from your garden, hand pick the weeds with their roots by using a root puller or a screwdriver. Weeds should be removed before seeds are produced, or the seeds will be spread around, and the weed problem will become worse.

Compost

Compost is organic matter decomposed or broken down by tiny organisms including worms, fungi, mites, and centipedes. As organic matter is broken down it produces a dark, nutrient-rich material that is a great natural fertilizer. In nature, all organic material is returned to the soil to add fertility, life, and nutrients for plants and other organisms. Composting speeds up the decomposition process by using organisms to transform organic matter into compost or humus. The composting process is complete when the material is dark and crumbly and smells fresh and earthy. This provides a balanced source of nutrients to enhance your soil, leading to healthier plants. Composting also helps reduce the amount of water your household sends to the landfill.

How to compost

Compost bins can be purchased or built. When building your own, use untreated wood. To start the composting process put 10 to 12 cm of household waste material in direct contact with the ground, then add 4 cm of fresh grass clippings or manure. The next layer is 2 cm of rich earth or completed compost and dust with dolomite limestone to keep rodents and other animals out of the bin.

How to worm compost

Worm composting can be done using a container made out of wood or plastic and that can be purchased or made at home. Worm composters can be used indoors or outdoors. Materials needed for a worm composter are: a wooden or plastic bin with a lid, 500 to 2000 red wiggler worms (*Eisenia foetida* or *Lumbricus rubellus*) depending on the size of your bin, bedding (shredded newspaper, leaves, straw, and other dead plants), two handfuls of sand or soil, and compostable wastes. Fill the compost bin with damp bedding until it is one thrid full. Add the two handfuls of sand or soil and then the worms. A general rule is to add one pound of worms for every pound of organic waste you will be composting each week. Bury the waste under the bedding and make sure large pieces of waste are cut up. The compost will be ready to harvest in two to three months.

What to compost

All uncooked, organic material can be composted. This includes fruits, vegetables, tea bags, coffee grounds (with filter paper), egg shells (make sure they are cleaned), grass clippings, leaves, weeds (make sure they are not laden with seeds), uncoloured paper products (newspaper, paper towels, napkins, brown bags), and sawdust. *Do not add dairy products, cooked foods, meat, fish, barbeque ash, or pet feces to your compost bin.*

Alternatives to Pesticides

All insects play a vital role in the food web. Although some insects are seen as pests, most insects help decomposition, are not harmful, and are an important food source for many birds. Only 2% of all insects are harmful to plants. But by using chemicals to rid your garden of these pests you will also kill other important insects. Pesticides are usually toxic to both the target and non-target species. Insecticides are known to kill bees, which are important for pollination, and ladybeetles, which are important in controlling pest populations. Pesticides also leave residues on plants and if birds or other animals eat these plants or dead bugs, it can stunt their growth, reduce reproduction rates, and cause death. Pesticides can also pollute the water that all organisms require to drink, bathe, and complete life cycles. Birds have been especially hard hit in the past 50 years and many other wildlife species have been adversely affected. Most household pesticide use is unnecessary as there are readily available alternatives. Alternatives to pesticides are very effective and environmentally friendly.

One of the easiest ways to control insect pests is by encouraging beneficial insects to take up residence in your garden. Beneficial insects such as ladybeetles, spined soldier bugs, and trichgramma wasps are natural enemies to agricultural pests. The easiest way to attract these insects to your yard is by planting species that attract them, such as tansy, yarrow, vetch, and dandelion. Pheromones, which are chemicals produced by female insects to attract males, can be purchased from nurseries. A last resort is to purchase these beneficial insects from a nursery and release them into your garden. If you choose to purchase insects from a nursery, ensure that they are native insects and not ones that are introduced from other continents.

Pests can also be physically removed from affected plants. This is done by hand picking infested leaves or pruning affected branches. Water sprayed directly at trees and shrubs is also effective at controlling some pests. If your insect infestation is severe, you can make your own insecticide by mixing 30 ml of soap flakes with one litre of water and dousing the infested leaves.

Companion planting is another way to naturally control garden pests. This is common in vegetable gardens, where one plant protects another from insect infestations: vegetables and flowers have different odours and root secretions that affect the activity of insects and growth of nearby plants. Marigolds and other members of the chrysanthemum family protect tomatoes, beans, and other plants from insects.

Watering

Water shortages are a growing concern throughout North America. In some cities, outdoor watering accounts for 35% of residential water use.

Native plants have evolved to withstand the weather conditions of the region. This means that when native plants are planted in your garden, watering requirements are minimal. Lawn dominated yards offer little habitat value to wildlife, but if you enjoy grass in your yard, find a variety that requires little water, pesticides, and fertilizers. If you must water your lawn, do so in the early morning to prevent the grass from burning, and only water once a week. A rain barrel is a great source of water for gardens and eliminates the dependence on tap water.

Balcony gardens require more water than yards, as they do not get natural rainfall. Water the plants before they begin to wilt, but do not set up a watering schedule. Assess the water requirements of your plants by looking at the surface of the soil daily in the summer and weekly in the winter. If the soil surface is dry all over, insert your fingertip. It is time to water if the soil is dry down to the length of your fingernail for a smaller pot and to your first joint in a larger pot. Water in the mornings and evenings, and avoid watering when there is frost on the plant. Do not dribble water daily. When it is time to water, fill the container with water to the lip (2.5 cm if the plant is planted correctly), and allow it to drain. Catch rainwater where you can on the balcony. If the leaves on your plant are yellow and wilted, with the tips turning brown and leaves are dropping off the entire height of the plant, you are watering too often or the container is not offering sufficient drainage. If the lower leaves are yellow and wilted with brown, dry edges and the lowest leaves are falling off you are not watering enough.

Feeding Your Plants

The following are signs to took for if you feel your plant might be nutrient starved:

- Nitrogen shortage — stunted growth, small, pale leaves and weak stems. To increase the amount of nitrogen, you can add a native nitrogen fixing plant. Large-leafed Lupine (*Lupinus polyphyllus*) will make nitrogen available to your other plants.
- Potash shortage — brown, brittle leaf edges, small flowers and low disease resistance
- Iron shortage — large yellow areas on leaves (most common in the newest foliage)
- Manganese shortage — yellow between the veins of the leaves (most common in oldest or lowest leaves).

Pruning

It is advantageous to remove the dead flowers from your plants for two reasons. It will prolong the display of other flowers on the plant, and it may induce a second flush of flowers later in the season. If, however, you want to propagate the species, let the flowers go to seed, and harvest them before they drop.

Early in the season, if you remove the growing point of the plant plus a small amount of the stem, you may stimulate bud production lower down the stem, creating side shoots and enhancing the bushiness of the plant. Generally, perennials need more pruning than shrubs and trees. Unless you are finding some species very straggly, or they are in the way, try to let them be.

Lawn Maintenance

Allow your lawn to grow tall and mow it infrequently. When you do mow your lawn, leave the clippings on the ground, as the nutrients will be returned to the soil. Water the lawn only when necessary, and water heavily once a week rather than several times a week. Adding a layer of mulch to you lawn also improves root growth and decreases the germination of weeds.

Gardening for Specific Wildlife

Birds

Birds are easily attracted to yards and balconies once food, water, and shelter have been provided. Common birds found in the Lower Mainland include Black-Capped Chickadee, Steller's Jay, Dark-Eyed Junco, American Robin, Northwestern Crow, Spotted Towhee, and House Finch.

By providing a source of food for birds, you can increase the number of species visiting your yard. These sources may either be natural or artificial. It is usually a good idea to provide both. Natural sources of food for birds are insects and seeds, nuts, and berries from native plants. Native plants also provide important shelter for many birds. Food can also be supplemented through feeders.

Feeders can help improve birds' survival rates. They are most important in late fall, winter, and early spring, when natural food sources are not as readily available. Feeders are most successful when they are placed in an area where there is a lot of vegetation for shelter and protection from predators. Birds are more likely to feed at feeders and plants when they feel comfortable. One feeder per yard or balcony is usually enough since most birds are territorial and extra feeders will just be wasted unless you have a large yard. If you choose to have more than one feeder in your yard, ensure that they are not concentrated in a small area. To enjoy your feeder, place it an area where you can see it from your house. It is also important to place the feeder where it can be easily accessed for refilling. Make sure feeders are cleaned once a week by removing old seeds and rinsing with vinegar and hot water.

Water is an effective tool to attract birds to yards and balconies. Clean, fresh water is often difficult for wildlife to find, and water is the only way to get some species to visit your yard. It is important for both drinking and bathing. Natural water sources include streams and ponds. There are many ways to provide water for birds in your yard if you are not lucky enough to have a natural source of water. The most attractive sources of water for birds are those that include movement, like dripping or flowing water.

A birdbath should have gently sloping sides to allow easy entry, and a rough surface is better than slippery plastic. The depth of the water should be no greater than 7.5 cm, and it is a good idea to provide rocks serving as perches for smaller birds. Birdbaths are most popular when they are placed near shrubbery for protection from predators. An attractive option that is compact enough for balconies is an overturned flower pot. The collection dish usually placed under the pot can be filled with water and placed on top of the overturned pot. You might want to paint your modified pots with non-toxic, waterproof paints. Unlike nesting boxes, bright colours are tolerated by birds around their water source. On a balcony you could create a layered effect, with a water pan slowly dripping into a birdbath, which overflows onto your plants.

It is very important to consider the cleanliness of the water you are providing. Never add chemicals of any kind to the water and change the water at least twice a week. In the winter, crack off any ice that forms, and birds will use the water source even if it is refreshingly chilled. After bathing, most birds perch to preen, which involves running their bill through their feathers to spread oil from a gland near the base of their tail over their feathers. Birdbaths should be cleaned regularly (once a week).

Shelter can be provided in many ways, depending on the birds you with to attract. Ground cover, shrubs, and brush piles are the perfect spot for ground nesting species such as juncos. Mature trees are nesting sites for many species that build their nests either in tree cavities or in the branches. Artificial habitat can also be provided for birds in your yard. Bird boxes are important to cavity nesting birds who have suffered a loss of habitat, such as chickadees, wrens, and swallows.

Building nest boxes is a fun community activity.

How to Build a Nesting Box

It is important that the entrance hole is large enough to allow the species you wish to attract in, but large enough to keep predatory birds out. Use untreated and unpainted wood, such as cedar, because it is natural, withstands weathering, provides insulation, and blends into the natural environment. Perches are not necessary for your bird box, since most native species that use nesting boxes do not require perches and they are used by predators looking for an easy meal.

It is nice to provide nesting material in a basket or mesh bag near your garden. Good nesting materials include short lengths of string and yarn, fleece, dried grass, small feathers and down, bulrushes, and cattails. Make sure not to place the nesting material in the nest box, as the birds will not use the box. Many birds need to construct their own nests as it helps them to become fertile and, in some species, is part of their courtship behaviour. Also, if they see things in a nest, they may think that it is already occupied.

The best time to clean out your bird box is during the fall season. It is very important to remove old eggs, dead nestlings, and old nesting materials,

Figure 15.1 Bird Box Plans

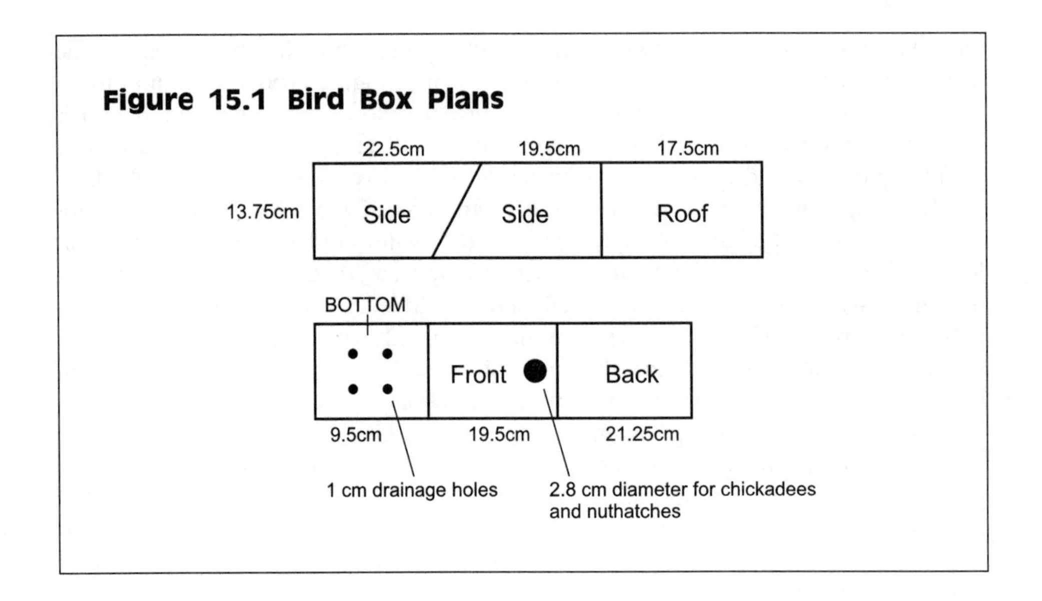

which contain parasites, dirt, and waste. Once the old nest has been removed from the box, boiling water can be poured through the box to ensure it is clean for next spring.

Figure 15.1 illustrates the dimensions of a bird box large enough for chickadees and nuthatches. Make sure the wood is 3/4" thick. The roof should be slanted and extend over the front and back. For ventilation, leave gaps between the roof and the front and back panels and drill four 3/8" holes in the floor of the box. Make the entrance hole no larger than 1 1/8" in diameter to keep out starlings and house sparrows, and roughen the surface of the wood just below the entrance on both sides of the wood.

In order to keep birds safe, bird boxes should be placed in an area that avoids direct exposure to the sun, prevailing winds, and driving rain. To protect birds from predators, boxes must be placed in the lower portion of a coniferous or orchard tree canopy or, similarly, in a region sheltered by shrubs. Furthermore, boxes that are mounted on trees should have a flight path that is unhampered by too many twigs and leaves. This will reduce potential hiding spots for predators. Where you place your bird box will determine the type of birds you attract. For example, chickadees and wrens prefer seclusion and shelter,

Bird Box Basics — A Place to Call Hole

So you've moved in and unpacked. You're all settled. But wait — did you remember the bird box? There are lots of friendly little animals that would love to come and entertain you in your new home. Birds are the most colourful, and they will sing for you too! You can help them out by building them a nest box.

To a bird, a nest box is like a hole in a tree. Lots of birds that are "hole-nesters" use them — chickadees, swallows, wrens, bluebirds, even Barn Owls and Wood Ducks. The size of the box and the diameter of the hole will determine what birds you will attract — and what undesirable hole-nesters like Starlings you can keep away.

whereas swallows and bluebirds like to have open access to their bird box. The best time to install your bird box is in February or March.

Hummingbirds

Hummingbirds are unique, and have habitat requirements different from those of other birds. They are some of the smallest warm-blooded vertebrates in the world. Their wings beat 80 times per second and, in forward flight, they can fly up to 95 km/hr. Hummingbirds' hearts beat up to 1,260 times per minute and they are the only birds capable of hovering and backward flight.

Their main food sources are nectar from plants and insects. One hummingbird can consume up to half its weight in food and eight times its weight in water each day. Their food requirements are similar to those of butterflies. They are attracted to yellow, pink, and red tubular flowers such as red-flowering currant and honeysuckle. There are over 130 plant species in North America that exhibit features that have been modified through time for foraging and pollination by hummingbirds. A plant list for a hummingbird garden is provided later in this chapter. It is important to plant in clusters around your yard rather than scatter plants throughout your yard, so hummingbirds have easy access to food. Hummingbirds also require dense shrubs as shelter at night. Water is another important consideration for hummingbirds, and can be provided through a shallow birdbath.

A box without a perch and with a 3 cm (1 1/8") hole will attract swallows and chickadees.

Feeders are a great way to supplement hummingbirds' food source. To attract hummingbirds, the feeder should be red. It is important to not add any food colouring, honey, or artificial sweetener to your feeder, as these are all toxic to hummingbirds. Feeders should be filled with four parts boiling water (boiling the water helps kill bacteria and reduce fluoride and chlorine) to one part sugar. Your feeder needs to be cleaned every three days, as bacteria and mold can grow in it and is toxic. Feeders should be hung in your garden near plants or on your balcony near a hanging basket or container. Since hummingbirds are very territorial, it better to space several small feeders throughout your yard if you wish to attract a number of hummers rather than have one large feeder.

Window Collisions

There are many options for reducing the chances of a visiting bird flying into your window. By placing birdbaths, nesting boxes, and feeders within one metre of windows, you will reduce the number of window collisions. With these targets close by, a bird will aim directly for them when landing and, if startled, will not be able to build up enough speed to injure itself in take off flight. Pulling your curtains closed in the day when you are out at work can also reduce the reflections created by your windows, which often confuses birds into thinking they can fly through the window, or that their reflection is a territorial rival to be charged. Turning off lights at night will avoid problems that we have with downtown office buildings. Many migrating birds get confused by the nighttime lighting in city high-rises, and either hit the windows or fly around the light until they fall from exhaustion. If you have repeated problems with window collisions, you can place a fine netting over your window to act as a "bird trampoline", or hang brightly coloured strips of paper on the outside of the window. Hawk silhouettes pasted on the insides of windows are not proven to work unless they are only 10 cm apart, which could block your view substantially.

Birds that have hit a window should be carefully picked up and placed in a cardboard box on top of your fridge and left alone (warm, dark and quiet). If you have pets keep them out of the room. Children will enjoy the encounter with your patient, but encourage them to leave the bird alone until it has recovered and they can watch it being released. After approximately two hours, the bird should be checked. If it is fluttering and moving normally, bring it outside at ground level and place the box with the lid off in a safe quiet place. Watch until it flies up into a tree or the safety of a shrub to make sure it does not have a run-in with the neighbourhood cats. If it is after 4 p.m., feed the bird with seed, peanuts or, if necessary, bread. Keep the bird overnight, and release it early the next morning. If the bird cannot fly, or seems injured, call your local wildlife rescue association.

Mammals

Mammals in your yard are very exciting and indicate that your yard is attractive to most wildlife. Common mammal species you can expect to attract to your yard include squirrels, raccoons, mice, voles, and bats. Larger mammals, such as deer and coyote, will only be attracted to your yard if you live close to large green space and your yard is not enclosed.

Most mammals prefer to find their own food and can be attracted to your yard by plants that produce berries, seeds, and nuts. Many mammals will also eat seeds off the ground that spill out of bird feeders. Ensure that your bird feeders are protected from mammals by using baffles or other guards and placing them away from the nearest point from which a squirrel could jump on to it. By sprinkling seeds, grain, corn, and nuts on the ground, you will provide adequate supplemental food for mammals.

As with birds and other wildlife, water is an essential component attracting mammals to your yard. They will use water in the form of puddles and birdbaths to drink and bathe. Brush and rock piles can provide small mammals with places to nest, den, and hide from predators. The value of such shelter is enhanced if there are internal spaces within the pile.

The goal with attracting mammals and other wildlife to your yard is to enjoy their presence, not to allow them to destroy your property or garden. Make sure areas under your porch, steps, deck, and patio are boarded up and have no way for animals to get in. Keep garbage in a secure container and try to keep it inside your garage until garbage day.

Bats

Bats make up approximately 20% of all mammal species, yet are probably the most misunderstood. They are very important in controlling insect populations: most bats can eat over 600 mosquitoes in one year. There are currently 16 species of bats in British Columbia. Eight of these species are endangered due to loss of habitat through forestry, urban development, and pollution of streams, ponds, and lakes.

How to Build a Bat Box

You can help create more habitat for bats by creating a bat house. Bat houses will not increase the likelihood that bats will move into your attic or wall. If bats liked your attic or walls, they would likely already be there.

Use a 12" × 72" board (Figure 15.2). Measure and outline the pattern on the board and then cut out the pieces. Roughen the inside of the front and back and both sides of the centre panels so the bats have a place to grip. You can use a wire brush to create deep grooves or staple thick plastic mesh to the appropriate sides. Line up the two side pieces with the back piece and hammer them together. The angle on the sides should be sloping downward. Attach the centre and front boards; each of the boards should have a 1" space between them. Attach the roof, which will angle down. Paint the house black to maxi-

Figure 15.2 Bat Box Plans

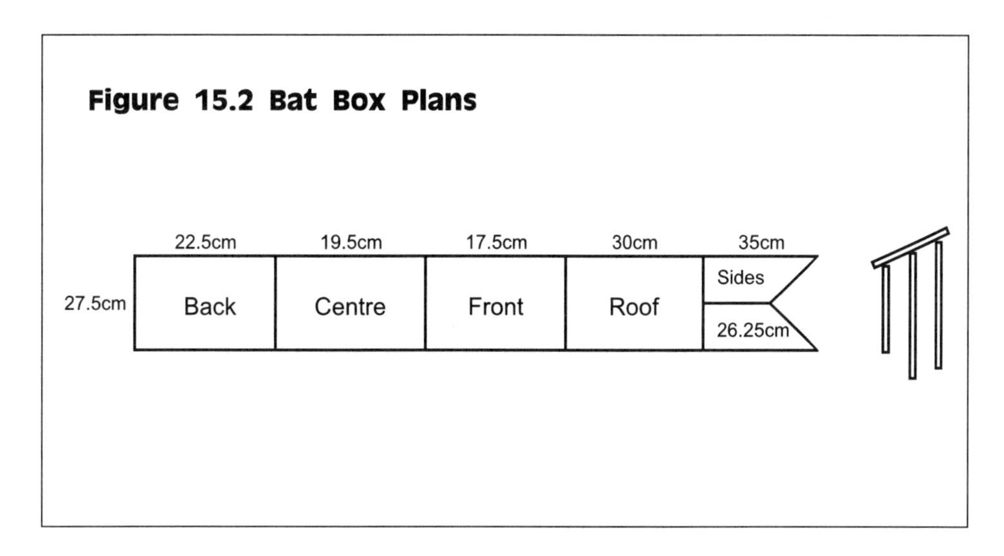

mize solar radiation and caulk the seams of the house, especially the roof, so it is watertight.

Hang the box on a pole or side of a building at least 10 feet off the ground, facing south or south-east where, it will get at least six hours of sun a day. Your bat house is more likely to be used if it is placed within 40 m of a stream, pond, or lake.

Evicting Bats from Your Attic

Although bats are important to our ecosystem, they can be a nuisance if they are living in your attic or walls. Because 50% of the bat species in BC are endangered, it is illegal to harm or kill them. Attics and other parts of buildings provide ideal roosting sites, and in most cases bats will not voluntarily leave. One thing you can do is evict them to a new location.

This bat box is an example of another way to support wildlife.

Bat evictions should occur in early spring or late summer when the flightless young are not present. First build and hang a bat box in a suitable location. At dusk, watch and see where the bats emerge from your house. Use polypropylene bird netting and hang a large enough piece over the emergence point, extending a foot below and to each side of the exit. Secure the net in place so that it hangs free an inch away from the building. This screen will allow the bats to exit, but not enter. Once all the bats have left and taken up residence in the bat house, you can board up the exit point.

Butterflies

Butterflies can bring beauty into your yard or balcony. There are approximately 15,000 species worldwide and 275 across Canada. Species common to the Lower Mainland of British Columbia include the woodland skipper, western tiger swallowtail, spring azure, painted lady, cabbage butterfly, and monarch. Fourteen of British Columbia's butterfly species are extirpated (no longer found in the region, but found elsewhere in its historical range), endangered (population is nearing extinction or extirpation), or threatened (likely to become endangered). The decline in butterfly populations is largely due to the use of pesticides and herbicides and the loss of habitat

Bat Box Basics — It's a Slot to Talk About

Did you know that one out of every five types of mammal is a bat? We don't see them much. They are creatures of the night after all. But they're out there, silently flying around, catching up to 600 mosquitos an hour, 3,000 a night.

There are many kinds of bats, but chances are that the little brown bat you may see in your neighbourhood is just that: it really is called the little brown bat. It is the most widespread and common bat in BC. It weighs about 10 grams, tops, about the same as a dime and two nickels. The change in your pocket probably weighs about as much as five of these bats.

After a hard night of acrobatics in the sky, a bat likes to find a cozy, warm spot to crawl into and sleep. In winter it needs a place to hibernate. Nothing is better than a loose piece of bark, rocky spaces in caves, or attics in old houses. There are not many of these around in, for example, Greater Vancouver, anymore. So, to help them out so they can help us control insects, we can make a bat box for them. The basic model (as illustrated by the schematic in Figure 15.2, and in the photo on the previous page) holds about 15 bats. The bats crawl into the slot from the opening below. It helps bats stay warm if you paint it black and put it about 10 feet above ground, where it catches the morning sun.

through urbanization. By planting a butterfly garden you can create a valuable habitat, and help decrease their risk of extinction. They prefer areas that are sunny, sheltered from the wind, have a good supply of larval food, have a reliable source of nectar, have a water supply, are free of chemicals, and have resting areas.

Butterflies can see a much broader colour spectrum than humans can — they see ultraviolet colours. Ultraviolet colours are very important to butterflies in their communication and food gathering: ultraviolet colours often show distinguishing features between males and females, and there are ultraviolet colours in many of the plants that butterflies get their nectar from.

Butterflies go through four stages of life — egg, caterpillar, pupa/chrysalis, and adult butterfly. Each stage is very different from the others and has different habitat and food requirements. When planting a butterfly garden, it is important to provide a variety of food for all stages of the life cycle. Caterpillars are the larval stage of butterflys' life cycle. They require much different food sources than adult butterflies. Larval plant food includes wildflowers, weeds, grasses, and trees (Stokes and Williams 1991). Common larval foods include aster, cottonwood, dogwood, lupine, stonecrop, and willow (Stokes and Williams 1991). Adult butterflies require plants that produce nectar to survive. Nectar providing plants should flower throughout the growing season so that butterflies are always able to find food. Adult butterflies are attracted to plants that provide a landing pad as well as easy access to the nectar, have strong, bright colours (orange, yellow, and purple), and have a strong fragrance. Clumping species together will attract more butterflies, as they are more attracted to large patches of bright, fragrant, and nutritious flowers than they are to plants that spread out.

Butterflies will also take advantage of feeding stations in your yard. A container of fresh or rotten fruit, such as bananas, apples, peaches, and grapes, in a sunny spot is very attractive. As soon as the food source has been discovered, the butterflies will return for more.

Butterflies often perform the behaviour of puddling. This is when they congregate and collect minerals from moist soil. Male butterflies are usually only found at puddling sites. They transfer nutrients consumed while puddling on to the female butterflies to help supply eggs with nutrients. To create a puddling site in your yard, excavate a small hole in damp soil or sink a container of sand to ground level. Old and clean paint trays, cake pans, and margarine containers work well; just make sure they are mostly filled with soil or sand and have only about a centimetre of water. Locate your puddling area in a sunny spot that is out of the wind and away from predators. It is also a good idea to place rocks around the perimeter to allow butterflies an area to sun themselves. Table salt can be added to the puddle to increase the mineral content.

A butterfly box.

Some species of butterflies spend the winter in this region as adults; these species include western painted ladies, red admirals, and mourning cloaks. These butterflies hibernate during the winter in tree crevices, under bark, in log piles, or in the nook of buildings.

Butterfly boxes are an excellent way to provide artificial habitat for butterflies and decorate your garden. A typical butterfly box is between 50 and 90 cm tall and 15 to 20 cm deep and wide. There are vertical slits in the front of the box that restrict access — the slits are 10 cm long and 1 cm wide. Place a strip of bark in the house to give the butterfly something to grasp onto. There also needs to be some way to open the box to put the bark inside and to clean it out. Boxes can be mounted in your garden on a tree or pole at least 1.5 m off the ground and near nectar producing plants.

Bees and Insects

Bees and insects are important pollinators for the plant in your garden. They are also an important food source for some of the larger wildlife that will use your habitat.

Bees and other insects are important to natural ecosystems because they pollinate flowers and plants. There are approximately 20,000 species of bee found throughout the world and 4,000 in North America. Unlike their relative the wasp, they only use their stinger in extreme cases for defence since, after they sting, they die. They are hairy and have large legs that have evolved to help collect pollen. Common bees found in urban areas of the west coast include mining (*Colletes cunicularis*), leafcutter (*Megachile spp*.), mason (*Osmia lignaria*), honey (*Apis mellifera*), and bumble (*Bombus terrestris*) bees.

To provide natural habitat for bees and insects, have flowering plants that produce a lot of pollen and nectar. Including ground cover, downed logs, woodpiles, stumps, and bark mulch around plantings also increases habitat and food for insects. Filling a shallow dish with water pebbles can provide water for bees and insects.

Mason Bees

Mason bees are native to North America, and are named because they use mud to seal their larvae filled nest. Mason bees are about two thirds the size of honeybees and it have a blue-black metallic colour. Due to its colour, the mason bee is sometimes mistaken for a fly. Females are slightly larger, and males have a white whisker-like moustache and larger antennae. They are solitary bees and do not nest in large colonies like honey or bumble bees. Unlike other bees they are not aggressive and do not usually sting unless they become aggravated or feel threatened. If they do sting, it is comparable to a mosquito bite.

Female mason bees are the pollinators and are able to pollinate in excess of 2,000 plants a day. In comparison, the honeybee can pollinate only 30 plants a day. Mason bees' home range is small and they will only search for pollen within 100 feet of their home. This means they will only pollinate flowers in your yard and, maybe, your neighbour's. Mason bees are generalists, meaning they are attracted to almost any plant that blooms in March, April or May.

Mason bees' life cycle begins in early spring. The adult males hatch and emerge from the nesting tubes first, as they are strategically laid towards the tube exit to protect the female larvae. Once all the bees hatch, mating occurs and the females lay their eggs two to three days later. Each female can lay up to 34 eggs. Each female prepare to lay eggs by collecting pollen and depositing it in the nesting tube as food for larvae. The egg is laid on the pollen ball and mud is collected to form a wall. The wall separates the nesting tube into chambers or cells. The female continues to lay eggs until the tube is full. Then it is closed off with a rough looking mud plug. This activity continues for four to six weeks until all the nesting tubes are full and then the adults die. Over the summer, the larvae develop into adults, and remain dormant until the next spring when it the cycle continues.

Since mason bees do not bore their own nests, they need to find one that is already made. They can nest in hollow stems or insect holes in trees or wood. Finding the correct size can sometimes be difficult. They require holes that are 5/16" diameter and at least 4" deep. You can put a bee box in your yard to help pollinate your flowers and provide habitat for mason bees.

Building a Mason Bee House

There are many styles of bee boxes using wood, plastic pop bottles, or ceramics. Mason bees will use a variety of nesting material. If you choose to use wood, make sure it has not been chemically treated. Mason bees are specific with their nesting requirement. The nesting tube size needs to be 5/16" diameter and a minimum of 4" deep. Each hole should be approximately 3/4" apart from its centres. There should be roof on the box for protection from sun and rain, and also ensures there is a slant on the roof to ensure proper drainage. In addition to using wood, bee boxes can be made out of 2-L pop bottles, coffee cans, and Styrofoam. Nesting tubes can be inserted inside these materials and can be made out of paper or plastic drinking straws. Remember, you can be creative with your bee box as long as the hole dimensions are correct.

When hanging your bee box, choose a location that provides shelter from direct wind, rain, and sunlight. Securely attach your bee box in a southern or southeastern position, allowing morning sun exposure. Since mason bees have a small home range, it is also important to ensure the box is hung near flowering plants.

Each nesting season, you should monitor your nest tubes for mites. One recommendation is to use newspaper or straws for nesting tubes, which can be discarded and replaced every year. The easiest cleaning method is to place a new bee box next to your old one, and the mason bees will colonize the clean house.

Mason bee nesting blocks are available to support the pollinators, such as the Blue Orchard Mason Bee.

Amphibians and Reptiles

Habitat for amphibians and reptiles has been drastically reduced by activities such as logging, urbanization, and drainage of wetlands. Clean sources of water are often difficult for these animals to find due to increased pollution and pesticide use. By helping create a freshwater source surrounded by vegetation, you can help increase their habitat.

Since amphibians and reptiles absorb water through their skin, they are not usually found too far from a source of water. This can include areas in or near ponds, marshes, streams, or rivers that often have muddy or rocky bottoms with

Bee Condos — It's Mainly Because of the Fruit

We aren't the only ones in the city living in condos. There is a little bee, blue black in colour and about two thirds the size of a honeybee, that likes the lifestyle too. All it needs is a little hole (normally a hole in a snag made by a beetle) and it will move in. The female puts a mud plug at the back of the hole and then constructs a few cells, each with an egg and some loads of nectar and pollen. The egg hatches, the young get stuffed with food, turn into adults, and then sleep off the feast until spring when they come out and get busy.

Have you ever thought of becoming a developer? Short on land or cash? Well, a bee condo may help you satisfy those cravings. A typical bee condo is just a block of wood with holes drilled into it. You can put it up next to your patio or balcony and wait for volunteers, or you could buy a few bees from most garden stores. Then watch. They're called mason bees for a good reason: they're pretty good at crafting with mud.

Mason bees are very handsome, safe (rarely sting and the sting is not bad if they do), loyal (they usually stay within a few hundred metres of your home), fun to watch and they will work hard for you — for free! They prefer visiting the blossoms of fruit trees, all of them — apples, cherries, plums, whatever you've got. So, sit back, watch them move into their condo, enjoy their company and then share your fruit harvest in the fall with friends — you'll have a bigger crop than usual.

a fair amount of aquatic vegetation to provide good hiding and resting places. Logs, rocks, and leaf litter not only provide good hiding places, but also attract a variety of insects that many amphibians like to eat. If you are more interested in attracting lizards and snakes, you many want to use loose, sandy soil around ponds with less leaf litter and more rocks.

A water garden in the middle of the lawn will have little wildlife habitat value. Make sure it is surrounded by tall grasses, ferns, shrubs, and tall perennials to provide shade and shelter. Be careful not to shade your water too much. Avoid removing leaf litter that accumulates around your pond, as it provides added habitat for ground dwelling insects that feed toads, frogs, salamanders, and birds. Add rock piles with cool depressions or dens underneath, amphibian houses, woodpiles, and sunning rocks around your pond to create habitat for amphibians and reptiles.

To create an amphibian house, use a clay pot that is 10 cm deep by 20 cm diameter. Cut a small arch as an entrance hole at the rim of the pot. Place the pot upside down within vegetation in a shady and secluded spot near a water source.

Native Plants

These tables are partial lists of plants native to the Pacific Northwest. Some plants for the northeast and the prairies are also mentioned, based in part on Knopf et al. (1995). Plant lists for other regions are readily available from gardening guides or through local natural history clubs.

Table 15.1 Trees

	Growing Conditions	Comments
Abies grandis Grand Fir	Dry to moist; shade.	Grows up to 80 m; habitat for owls, woodpeckers, toads, frogs, and salamanders.
Acer circinatum Vine Maple	Moist; shade tolerant.	Grows up to 7 m; seeds are a food source for birds and small mammals.
Acer macrophyllum Bigleaf Maple	Dry to moist; sun to partial shade.	Grows up to 35 m; seeds are eaten by birds and small mammals.
Alnus rubra Red Alder	Moist; full sun to partial shade.	Grows up to 25 m; attracts chickadees and other birds.
Cornus nuttallii Pacific Dogwood	Moist; partial to full shade.	Grows up to 15 m; birds are attracted to bright red berries.
Crataegus douglasii Black Hawthorn	Moist; full sun.	Grows up to 10 m; source of food for birds in winter; thicket provides nesting sites for birds and small mammals.
Malus fusca Pacific Crab Apple	Moist to wet; full sun.	Grows between 2 and 12 m; provides food from July to October.
Prunus emarginata Bitter Cherry	Moist; full sun to partial shade.	Grows between 2 and 15 m; bright red cherries eaten in early fall.
Pseudotsuga menziesii spp. *Menziesii* Douglas-fir	Dry to moist; full sun.	Grows up to 80 m; food source for songbirds and small mammals.
Thuja plicata Western Redcedar	Moist to wet; shade.	Grows up to 60 m; seeds eaten by songbirds.

Some suitable native trees for other regions of North America include:

- Northeast: eastern red cedar (*Juniperus americana*), Canadian hemlock (*Tsuga canadensis*), eastern white pine (*Pinus strobus*), red maple (*Acer rubra*), sugar maple (*Acer sacchrum*), American beech (*Fagus grandifolia*), paper birch (*Betula papyrifera*), white oak (*Quercus alba*), pin cherry (*Prunus pensylvanica*) and American mountain ash (*Sorbus Americana*)
- Central prairies and Great Plains: eastern cottonwood (*Populus deltoides*), bur oak (*Quercus macropcarpa*), downy hawthorne (*Cratagus mollis*), smooth sumac (*Rhus glabra*) and green ash (*Fraxinus pensylvanica*).

Theme/Demonstration Gardens

Theme gardens are an excellent way to attract certain wildlife species to your yard and are also a way to help you learn more about the plants you are using. These gardens are fun to design and can be planted anywhere: backyards, balconies, and containers. The plants listed are native to coastal British Columbia.

Table 15.2 Shrubs

	Growing Conditions	Comments
Amelanchier alnifolia Saskatoon	Dry to moist; full sun.	Grows up to 5 m; deciduous; provides birds with food through the fall.
Cornus stolonifera Red-Osier Dogwood	Moist to wet; full sun to shade.	Grows up to 4 m; deciduous; fall and winter food source for birds.
Corylus cornuta var. *californica* Beaked Hazelnut	Moist; full sun to partial shade.	Grows between 1 and 4 m; deciduous; nuts are an autumn food source.
Gaultheria shallon Salal	Dry to wet; full sun to partial shade.	Grows up to 5 m; evergreen; berries are food for birds and deer browse on leaves throughout the year.
Lonicera ciliosa Orange Honeysuckle	Moist; partial sun to shade.	Climbing plant up to 6m; deciduous; attracts hummingbirds and butterflies.
Lonicera involucrate Black Twinberry	Moist to wet; shade.	Grows up to 3 m; deciduous; yellow flowers produce nectar for hummingbirds; berries provide food for birds.
Mahonia aquifolium Tall Oregon Grape	Dry to drought tolerant; full sun to partial shade.	Grows up to 6 m; berries attract birds.
Oemleria cerasiformis Indian Plum	Dry to moist; full sun to partial sun.	Grows up to 5 m; deciduous; fruits attract birds.
Physocarpus capitatus Pacific Ninebark	Wet; full sun.	Grows up to 4 m; deciduous; flowers are a source of nectar.
Ribes sanguineum Red Flowering Currant	Dry; full sun to partial shade.	Grows up to 3 m; deciduous; flowers attract hummingbirds and butterflies.
Rosa nutkana *Nootka Rose*	Dry to moist; full sun.	Grows up to 3m; deciduous; flowers attract bees; rosehips last into the winter.
Rubus parviflorus Thimbleberry	Dry to moist; full sun.	Grows up to 3 m; deciduous; berries provide birds with food.
Rubus spectabilis Salmonberry	Moist to wet; sun to shade.	Grows up to 4 m; deciduous; attracts songbirds and hummingbirds.
Salix hookeriana Hooker's Willow	Wet; full sun to partial shade.	Grows up to 6 m; deciduous; pollen is food source for many insects, including bees and moths.
Sambucus racemosa ssp. *pubens* var. *arborescens* Red Elderberry	Dry to moist; full sun to partial shade.	Grows up to 6m; deciduous; flowers provide nectar for hummingbirds and butterflies; berries are food for birds.
Spiraea douglasii Hardhack	Moist to wet; full sun.	Grows up to 2 m; deciduous; flowers attract bees.
Symphoricarpos albus Common Snowberry	Dry to moist; full sun to partial shade.	Grows up to 2 m; deciduous; flowers attract bees and hummingbirds; berries remain until December.
Vaccinium ovatum Evergreen Huckleberry	Dry to moist; full sun to shade.	Grows between 1 and 4 m; evergreen; flowers attract hummingbirds; berries remain until early winter and are food for birds.
Vaccinium parvifolium Red Huckleberry	Dry to moist; full sun to partial shade.	Grows up to 4 m; deciduous; berries are eaten by birds throughout the summer.

Table 15.2 Shrubs Cont.

Viburnum edule Highbush Cranberry	Moist; full sun to partial shade.	Grows between 0.5 and 3.5 m; deciduous; fruit remains through the winter and is an important source for over-wintering birds.

Some native shrubs suitable for other regions of North America include:

- Eastern: highbush blueberry (*Vaccinium corymbosum*), bayberry (*Myrica pensylvanica*), red osier dogwood, red chokeberry (*Pyrus arbutifolia*), Virginia creeper (*Parthenocissus quinquefolia*), wild grape (*Vitis spp.*) and white elderberry (*Sambus canadensis*)
- Central prairies and Great Plains: Wood's rose (*Rosa woodsi*) and coyote willow (*Salix exigua*).

Table 15.3 Ground Cover

	Growing Conditions	Comments
Arctostaphylos uva-ursi Kinnikinnick	Dry; full sun.	Grows up to 0.2 m; evergreen; attracts hummingbirds, and berries provide food for birds.
Rubus ursinus Trailing Blackberry	Dry; full sun.	Trailing to 5 m long and 0.5m high; deciduous; berries attract birds in late summer; provides food for deer in winter.

Some native ground cover suitable for other regions of North America include:

- Eastern: Kinnikinnick (bearberry)

Table 15.4 Ferns

	Growing Conditions	Comments
Polystichum munitum Sword Fern	Moist; partial to full shade.	Grows to 1.5 m.
Athyrium felix-femina Lady Fern	Moist to wet; partial to full shade.	Grows to 2 m.
Blechnum spicant Deer Fern	Moist to wet; partial to full shade.	Grows between 0.2 and 0.8 m; winter food source for deer.
Pteridium aquilinum Bracken Fern	Dry to wet; partial to full shade.	Grows to 3 m.

Some native ferns suitable for other regions of North America include:

- Eastern: lady fern, bracken fern, ostrich fern (*Matteuccia struthiopteris*) and grape fern (*Botrychium spp.*)
- Central prairies and Great Plains: lady fern and bracken fern

Table 15.5 Perennials

	Growing Conditions	Comments
Achilela millefolium Yarrow	Dry to moist; full sun to partial shade.	Grows to 1 m; blooms June to July; attracts butterflies and hummingbirds.
Anaphalis margariacea Pearly Everlasting	Dry; full sun to partial shade.	Grows to 1 m; blooms July to August; summer browse for deer.
Aquilegia formosa Red Columbine	Moist; full sun to partial shade.	Grows to 1 m; blooms May to June; flowers attract butterflies and hummingbirds.
Dicentra formosa Pacific Bleeding Heart	Moist; shade.	Grows to 0.5 m; blooms May to June; attracts hummingbirds; seeds are spread by ants.
Epilobium angustifolium Fireweed	Dry to moist; full sun.	Grows to 1 m; blooms June to July; attracts hummingbirds and bees; fixes nitrogen in soil.
Heracleum lanatum Cow-parsnip	Dry to moist; full sun and partial shade.	Grows to 3 m; blooms May to June; attracts butterflies and bees.
Lupinus ssp. Lupine	Dry to wet; full sun to partial shade.	Grows to 1.5 m; blooms early spring; attracts bees and hummingbirds; fixes nitrogen in soil.
Sedum spathulifolium Broad-leaved stonecrop	Dry; full sun.	Grows to 0.2 m; blooms May to June.
Solidago canadensis Canada Goldenrod	Dry to moist; full sun to partial shade.	Grows to 1.5 m; blooms May to June; attracts insects.
Stachys cooleyae Cooley's hedgenettle	Moist; full sun to partial shade.	Grows between 0.7 and 1.5 m; flowers attractive to hummingbirds.

Some native perennials suitable for other regions of North America include:

- Eastern: black-eyed susan (Rudbeckia hirta), New England aster (*Aster novae-angliae*), columbine (*Aquilegia canadensis*), bunchberry (*Cornus canadensis*) and rose mallow (*Hibiscus moscheutos*)
- Central prairies and Great Plains: day flower (*Commelina dianthifolia*), rose verbena (*Verbena canadensis*), blue sage (*Salvia azurea*), goldenrods (*Solidago* spp.), wild bergamot (*Monarda fistulosa*), Maximilian's sunflower (*Helianthus maximiliani*), big bluestem (*Andropogon gerardi*) and blue grama (*Bouteloua gracilis*)

Table 15.6 Butterfly and Hummingbird Garden

Common name	Scientific name
Vine Maple	*Acer circinatum*
Black Twinberry	*Lonicera involucrate*
Common Snowberry	*Symphoricarpos albus*
Evergreen Huckleberry	*Vaccinium ovatum*
Mock-Orange	*Philadelphus lewisii*
Nootka Rose	*Rosa Nutkana*
Oceanspray	*Holodiscus discolor*
Red Elderberry	*Sambucus racemosa* ssp. *pubens*
Red-Flowering Currant	*Ribes sanguineum*
Salmonberry	*Rubus spectabilis*
Timbleberry	*Rubus parviflorus*
Kinnikinnick	*Arctostaphylos uva-ursi*
Orange Honeysuckle	*Lonicera ciliosa*
Canada Goldenrod	*Solidago Canadensis*
Cow-Parsnip	*Heracleum lanatum*
Arctic Lupine	*Lupins arctiucs*
Fireweed	*Epilobium angustifolium*
Large-Leaved Lupine	*Lupinus polyphyllus*
Nodding Onion	*Allium cernuum*
Pacific Bleeding Heart	*Decentra Formosa*
Pearly Everlasting	*Anaphalis margarticacea*
Red Columbine	*Aquilegia Formosa*
White Fawn Lily	*Erythronium oregonum*
Yarrow	*Achillea millefolium*

Butterflies for Peace — Tranquility in the Home

There is little in the natural world that matches the tranquility and innocence of a butterfly. You're sitting outside enjoying a drink, a western tiger swallowtail drifts by with its flashes of yellow and black. It is living art. You pause. It works its magic, silently anointing your yard or balcony with its wings, and moves on. Evil spirits are banished. How can you make this happen more often? How can you attract other backyard and balcony regulars like the woodland skipper or the painted lady or the spring azure to entertain you and your friends?

Butterflies like lots of trees, shrubs and wildflowers. Mock orange, Nootka rose, red flowering currant, salmonberry, red columbine — almost any plant with a colourful, aromatic flower is fine. A popular choice is the Buddleia or butterflybush which, mercifully, is easy to grow and attracts many different kinds of butterflies. Or, a bouquet in a hanging basket. Butterflies like plants whose flowers provide a landing pad as well as access to nectar. They are attracted to flowers with strong colours like orange, yellow and purple, and that have a strong fragrance.

If you are handy with wood, you could build a butterfly box. Staple some bark or moss on it and it will make a nice garden ornament. Butterflies can use it to hibernate over the winter. They won't snore. They'll be quiet and tranquil.

Table 15.7 Bug Garden

Pacific Willow	*Salix lucida* ssp. *lasiandra*
Common Snowberry	*Symphoricarpos albus*
Nootka Rose	*Rosa nutkana*
Oceanspray	*Holodiscus discolor*
Red Elderberry	*Sambucus racemosa* ssp. *pubens*
Arctic Lupine	*Lupins arctiucs*
Canada Goldenrod	*Solidago Canadensis*
Large-Leaved Lupine	*Lupinus polyphyllus*
Nodding Onion	*Allium cernuum*
Pearly Everlasting	*Anaphalis margarticacea*
Red Columbine	*Aquilegia Formosa*

Table 15.8 Wildflower Mix

		Colour
Arctic Lupine	*Lupinus arcticus*	Purple
Canada Goldenrod	*Solidago Canadensis*	Yellow
Cow-Parsnip	*Heracleum lanatum*	White
Fireweed	*Epilobium angustifolium*	Pink
Fingecup	*Tellima gradniflora*	White
Large-Leaved Lupine	*Lupinus polyphyllus*	Purple
Nodding Onion	*Allium cernuum*	Lavender
Pearly Everlasting	*Anaphalis margaritacea*	White

Table 15.9 Summer Bird Food

Cascara	*Rhamnus purshiana*
Black Twinberry	*Lonicera involucrate*
Evergreen Huckleberry	*Vaccinium ovatum*
Indian-Plum	*Oemleria cerasiformis*
Red Elderberry	*Sambucus racemosa* ssp. *pubens*
Salmonberry	*Rubus spectabilis*
Tall Oregon-Grape	*Mahonia aquifolium*
Thimbleberry	*Rubus pariflorus*
Kinnickinnick	*Artostaphylos uva-ursi*
Trailing Blackberry	*Rubus ursinus*

Indian plum is one of the first shrubs to come into leaf in spring and its berries offer excellent food for wildlife.

Table 15.10 Fall and Winter Bird Food

Big Leaf Maple	*Acer macrophyllum*
Douglas-Fir	*Pseudotsuga menziesii* ssp. *menziesii*
Sitka Spruce	*Picea sitchensis*
Western Hemlock	*Tsuga heterophylla*
Western Redcedar	*Thujs plicata*
Black Hawthorn	*Cratageus douglasii*
Pacific Crab Apple	*Malus fusca*
Beaked Hazelnut	*Corylus cornuta*
Black Raspberry	*Rubus leucodermis*
Common Snowberry	*Symphoricarpos albus*
Evergreen Huckleberry	*Vaccimium ovatum*
Red-Osier Dogwood	*Cornus stonoifer*
Saskatoon	*Amerlanchier alnifolia*
Orange Honeysuckle	*Lonicera ciliosa*
Trailing Blackberry	*Rubus ursinus*

Table 15.11 Ethnobotany Garden

Wester Redcedar	*Thuja plicata*
Vine Maple	*Acer circinatum*
Beaked Hazelnut	*Corylus cornuta*
Black Gooseberry	*Ribes lacustre*
Evergreen Huckleberry	*Vaccinium ovatum*
Mock-Orange	*Philadlphus lewisii*
Nootka Rose	*Rosa nutkana*
Oceanspray	*Holiduscus discolor*
Pacific Ninebark	*Physocarpus capitatus*
Red Elderberry	*Sambucus racemosa* ssp. *pubens*
Red Huckleberry	*Vaccinium parvifolium*
Saskatoon	*Amelanchier alnifolia*
Tall Oregon-Grape	*Mahonia aquifolium*
Orange Honeysuckle	*Lonicera ciliosa*
Trailing Blackberry	*Rubus ursinus*
Cow-Parsnip	*Heracleum lanatum*
Goat's beard	*Arunucus dioicus*
Nodding Onion	*Allium cernuum*
Red Columbine	*Aquilegia Formosa*
Skunk Cabbage	*Lysichiton americanum*
Wild Ginger	*Asarum caudatum*
Yarrow	*Achillea millefolium*
Lady Fern	*Athyrium filix-femina*
Sword Fern	*Polystichum munitum*

Table 15.12 Food Garden

Beaked Hazelnut	*Corylus cornuta*
Black Gooseberry	*Ribes lacustre*
Evergreen Huckleberry	*Vaccinium ovatum*
Indian-Plum	*Oemleria cerasiformis*
Labrador Tea	*Ledum groenlandicum*
Nootka Rose	*Rosa nutkana*
Red Elderberry	*Sambucus racemosa* ssp. *pubens*
Red Huckleberry	*Vaccinium parvifolium*
Salal	*Gaultheria shallon*
Salmonberry	*Rubus spectabilis*
Soapberry	*Shepherdia canadensis*
Tall Oregon-Grape	*Mahonia aquifolium*
Thimbleberry	*Rubus parviflorus*
Trailing Blackberry	*Rubus ursinus*
Cow-Parsnip	*Heracleum lanatum*
Nodding Onion	*Allium cernuum*
Common Horsetail	*Equisetum arvense*
Sword Fern	*Polystichum munitum*
Lady Fern	*Athyrium felix-femina*

Table 15.13 Alpine Garden

Thirt Seapink	*Armeria maritima*
Common Harebell	*Campanula notundifolia*
Scouler's Harebell	*Campanula scouleri*
Lance-fruited Draba	*Draba lonchocarpa*
Northern Geranium	*Geranium erianthum*
Broad-leaved Stonecrop	*Sedum spathulifolium*
Roseroot	*Sedum integrifolium*
Oregon Stonecrop	*Sedum oreganum*
Spreading Stonecrop	*Sedum divegens*
Lance-leafed Stonecrop	*Sedum lanceolatum*
Spreading Phlox	*Phlox diuffusa*
Yellow Wood Violet	*Viola glabella*

Table 15.14 Xeriscaping Garden

Bigleaf Maple	*Acer macrophyllym*
Douglas Maple	*Acer glabrum*
Douglas-Fir	*Pseudotsuga menziesii* ssp. *menziesii*
Western Flowering Dogwood	*Cornus nuttallii*
Baldhip Rose	*Rosa gymnocarpa*
Common Snowberry	*Symphoricarpos albus*
Indian-Plum	*Oemleria cerasiformis*
Mock-Orange	*Philadelphus lewisii*
Nootka Rose	*Rosa nutkana*
Oceanspray	*Holodiscus discolor*
Red-Flowering Currant	*Ribes sanguineum*
Redstem Ceanothus	*Ceanothus sanguineus*
Salmonberry	*Rubus spectabilis*
Saskatoon	*Amelanchier alnifolia*
Soapberry	*Shepherdia canadensis*
Tall Oregon-Grape	*Mahonia aquifolium*
Thimbleberry	*Rubus parviflorus*
Kinnikinnick	*Arctostaphylos uva-ursi*
Trainling Blackberry	*Rubus ursinus*
Arctic Lupine	*Lupinus arcticus*
Common Red Paintbrush	*Castilleja miniata*
Cut-leaf anemone	*Anemone multifida*
Harsh Paintbrush	*Castilleja hispida*
Harvest brodiaea	*Brodiaea coronaria*
Indian consumption plant	*Lomatium nudicaule*
Nodding Onion	*Allium cernuum*
Pearly Everlasting	*Anaphalis margaritacea*
Satin-Flower	*Sisyrinchium douglasii*
White Fawn Lily	*Erythronium oregonum*
Yarrow	*Achillea millefolium*

Indian plum is one of the first shrubs to come into leaf in spring and its berries offer excellent food for wildlife.

Don't limit yourself to these suggestions. Be creative! Other demonstration garden ideas include wildlife habitat for amphibians, reptiles, small mammals, and ground nesting birds. Plant lists for demonstration gardens from Eakins et al. 1999.

16

Agents and Tools for Protecting and Managing Urban Biodiversity

Urban biodiversity can be protected at many different levels. Individual action and the efforts of community groups and nongovernmental organizations have an important role to play. This is especially true about adopting lifestyles that are less harmful to the environment. Small gains from individual action can lead to big environmental benefits.

However, large-scale solutions usually require the attention of business and various levels of government — municipal, regional, provincial or state, and national or federal. These best management practices, policy, legislated solutions and stewardship programs can assist individual efforts to accomplish things beyond the scope of individual action (GVRD 2003). Examples include:

- regulatory tools that may limit impervious surface, control stormwater runoff, create parks or establish curbside recycling programs
- alternative development standards that can address parking lots or roads, construction site management practices such as amending soils and sediment control
- habitat enhancements, such as biofiltration ponds and using native plants in landcaping
- farmland stewardship that may include cover crops, grassland set asides and hedgerows

The combination of individual action, business practices and government legislation that most effectively conserves urban biodiversity.

One major difference between conserving or enhancing urban biodiversity compared to a wilderness situation is that it needs to be sensitive to the human context. Although parks and other natural settings are appreciated by most people, there are certain constraints that influence whether a green space will be valued and endure. Kaplan and Kaplan (1995) determined that nature scenes have low complexity and high preference compared to streetscapes of buildings that have a high degree of complexity but low preference. However, at the same time, green spaces need to be legible and mysterious; there is a need for both understanding and exploration. Legibility requires an easily understood landscape with landmarks and meaningful aids to achieve legibility (perhaps signs). Most preferred nature scenes are those with mystery and with a promise of further information given the opportunity to walk deeper into the green space.

The work of Kaplan and Kaplan indicates that creating habitat to improve urban biodiversity should also consider the legibility and mystery of the habitat. The long-term survival of any efforts to improve urban biodiversity depends upon the political support, and economic resources and stewardship of the people who live there. In addition to legibility and mystery, people need to feel comfortable with the green space. This involves feeling safe.

People prefer a landscape with recognizable features. An open space such as a lawn does not rate well. Neither does a dense tangle of understorey dominating the foreground of a scene. People who live in an area have a greater ability to discern recognizable features than visitors or tourists. They will tolerate more "mess" in the vegetation of a landscape in their neighbourhood cthan they will in other situations where the green space is a park that serves visitors that do not live in the immediate area. There are also age differences in preferences for green space. Teenagers, for example, generally do not prefer natural settings as much as children and adults.

The agents and tools used to manage and protect urban biodiversity are constantly challenged by the psychological characteristics of the people sharing the same space.

Individual Action — Environmental Citizenship and Environmental Stewardship

What does it mean to be a citizen? Basically it means to become informed on an issue (education) and then act on the basis of that information. With regards to urban biodiversity and its protection, it involves learning more about its composition and threats and letting others know about it. It involves learning more about what can be done, the barriers to doing something and problem solving to overcome those barriers.

Education is closely tied to "awareness", and is the easier part of environmental citizenship. The acting part is more difficult. People frequently have the information and are very aware of the threats to urban biodiversity, but do not act. In such situations, even more information does not lead to action; it is not a matter of quantity: there is a quality gap between information and action. People need to be motivated to care, to value urban biodiversity. More scientific information doesn't help, but compelling photographs, murals and writings can emotionally move people to action.

A steward is someone who looks after something that does not belong to them, usually a piece of land. Someone could choose to become an environmental steward as part of their contribution to environmental citizenship. The Environmental Streamkeepers Program of British Columbia is an excellent example of stewardship. These groups are tied to watersheds as their land base that they look after (e.g. Hyde Creek Streamkeepers, Maple Creek Streamkeepers, Scott-Hoy Creek Streamkeepers). They plant native vegetation, clean refuse, stabilize stream banks and may even raise salmon fry to release into the streams.

Business Best Management Practices

Businesses can adopt environmentally friendly practices to reduce pollution, energy consumption and waste production. They can receive certification for these practices from organizations such as ISO 14000 (www.iso14000.com) that lets their clients know of their commitment to the environment. Included in the evaluation is a broad definition of the environment to extend from within the organization itself to the global system. It examines:

- how the organization's activities, products or services interact with and impact on the environment
- management systems, such as organizational structure, planning activities and processes
- auditing procedures of the management system
- the existence and content of an environment policy
- performance targets to meet environmental objectives

Artistic representations of nature as in a mural can inspire people in situations where living material is not possible.

The most direct and immediate benefit to the natural environment in an urban area would be through landscaping and building design. Landscape Architects can design a number of natural areas suitable for building grounds. They would be similar to any number of the habitats described in earlier chapters.

A voluntary, consensus-based national standard for environmental building design exists through Leadership in Energy and Environmental Design (www.leedbuilding.org). LEED standards were developed by the US Green Building Council, whose members represent all segments of the building industry. The "green building" concept applies to high-performance sustainable buildings, and includes:

- reusing existing buildings (maintaining a percentage of shell and nonshell components)

- reusing salvaged or refurbished materials
- using rapidly renewing materials, usually resources that can be planted and harvested within a 10-year period
- using lumber with sustainable forestry certification
- using low-emitting materials indoors with regard to adhesives, sealants, paints, carpets and composite wood.

Government Action — Municipal

Cities are a mosaic of development resulting from historical and present-day priorities in land use practices. Ideally, the conservation of biodiversity would be a primary concern, and development would occur around environmentally sensitive areas, and would be sensitive to the need to reduce impacts on these sites. However, that is not the case. Instead, development builds from the current infrastructure, perpetuating environmentally damaging practices.

Although municipalities are constrained by their existing pattern of development, much can be done to guide future development and conserve biodiversity through official community plans, bylaws and other tools.

An official community plan can provide for rapid transit that will reduce vehicle traffic and require fewer roads.

Official Community Plans

Municipalities produce Official Community Plans (OCPs) that contain goals, objects and specific designations that deal with many issues affecting urban biodiversity. They describe how the community will approach land use, development and environmental issues in terms of its attitude and commitments.

For example, if the goal is to protect fish and wildlife habitat, the OCP and the policies derived from it will contain objectives that (Stewardship Series 1994):

- Conserve leave areas adjacent to watercourses
- Control soil erosion in runoff
- Control instream work, construction and diversions
- Prevent discharge of pollutants into watercourses

Urban planning has its greatest impact on the environment through the land use it determines. Longer commutes between home and work mean more roads, habitat loss, air pollution and urban runoff. Sustainable communities are those that address long term environmental impacts in their design and function.

One charette that examined ways to create a sustainable community in a planned development at East Clayton in Surrey, BC, produced six simple principles for sustainable communities (Condon 1996). These are:

1. Locate different dwelling types in the same neighbourhood and even on the same street. Larger homes could be integrated with duplexes, apartments and secondary rental suites. This would accommodate diverse housing needs of people from a wide range of income levels.
2. Provide buildings that present a friendly face to the street. This promotes a people friendly streetscape, encouraging pedestrians and more walking between destinations.
3. Transit and shops should be within a 5-minute walking distance. In addition to encouraging walking on shopping trips, this encourages a minimum density for a viable transit system.
4. Provide an interconnected street system. Having a grid pattern establishes minimum walking distances between destinations instead of having to go around cul-de-sacs or other obstacles.
5. Provide lighter, greener, cheaper, smarter infrastructure. This reduces the amount of pavement needed to support the lifestyle of the typical suburban dweller which, at the moment, is four times more pavement than the average urban dweller.
6. Provide natural drainage systems. Neighbourhoods should be drained in the same way as the original forests or grasslands were drained. This would help maintain the livelihood of streams.

All of these principles help to maintain the functioning of natural ecosystems and serve to maximize the biodiversity that can be conserved in the light of urban development.

Environmental policy is embedded in a broader context of sustainability that involves long-term political goals addressing social issues. The priority of maintaining the natural environment and biodiversity loses legitimacy to concerns over economic factors such as marketability, efficiency

Official community plans need to ensure adequate green space for people and wildlife.

and competitiveness. An agenda of such ecological modernization is widely applied as, for example, in dealing with contaminated soil and the Don River in Toronto, and air pollution and the Los Angeles River in Los Angeles (Keil and Desfor 2003). In dealing with contaminated sites and air pollution, these cities chose to support market driven regulation measures rather than just strictly imposed government guidelines, which during times of economic downturn were too costly. In dealing with their rivers, both cities adopted an approach of a series of small steps to restore ecological integrity.

With ecological modernization, urban planning and economics consider societal, economic, technological and social change, trying to create win-win situations. Global initiatives such as Agenda 21 have provided a general framework for municipalities to both improve local environments and further urban growth.

Bylaws

Municipal bylaws are one of the more powerful tools for protecting urban biodiversity. A few examples are given here (Stewardship Series 1994):

Zoning Bylaws

1. Setbacks
   - establish minimum leave areas to keep development away from the top of a stream bank for a certain distance
2. Limited Disturbance Clauses
   - limit activities within the setbacks: for example, restricting soil disturbance
3. Environmentally Sensitive Areas (ESAs)
   - permitting uses to avoid locating developments next to a stream that might degrade the habitat for fish
   - restrict access to controlled points by using fencing
   - have larger parcel sizes adjacent to ESAs to allow for adequate setbacks

Tree Protection Bylaws

A number of municipalities now limit the number of trees beyond a certain size that can removed from a property. Large and historically significant trees are viewed as community assets. They have significance beyond the property on which they are located. Such trees may be given strong protection from destruction. If they must be removed, there are measures to compensate for the loss, such as planting a number of replacement trees.

Rezoning

When land is rezoned, either at the initiation of the landowner or local government, it provides an opportunity to include habitat protection measures in the future use of the site. There may be some tax savings for the landowner as a result, or the landowner may be able to transfer development rights by rezoning two or more parcels at the same time

Incentive Zoning

Incentive Zoning of areas of wildlife habitat can allow flexibility that will protect biodiversity.

1. Density bonusing
   - A developer may be apply for a density bonus (put in extra units) if they build in a certain area in exchange for the developer protecting an ESA
2. Comprehensive development zones
   - Working within definitions contained in the OCP, the municipality creates customized zoning regulations with permitted uses, densities and desired amenities to deal with a complex site.
3. Development variance permits
   - Variances can be granted to normal regulations such that, for example, the setback of a house from a street can be reduced in exchange for protecting a stream corridor at the back of the property

Land Tenure

There are several options available to protect land when it is being developed. These will determine ownership of the land and what building can occur, not only at the present but into the future. These options include (Stewardship Series 1994):

1. Dedication
   - The management and maintenance of land can be dedicated to a local government
2. Covenants
   - A conservation covenant can be applied to a property that restricts public access in an ESA, or redevelopment of a private property
3. Voluntary Stewardship
   - Includes tools such as contracts, agreements, memoranda of understanding, leases and trusts

Parks and Open Space

Municipalities may require that a landowner who subdivides their property dedicate a certain percentage of their land for parks or cash in lieu. Similarly there may be provision for the mandatory dedication of wetlands and shorelands as occurs in Alberta, Saskatchewan and New Zealand (Sandborn 1997).

Municipalities also assume responsibility for the management of many parks and open spaces. Acquiring property for public green space and developing and implementing appropriate management plans are increasingly important activities for municipalities as they strive to make their communities more liveable. For example, municipalities are concerned about adequate recruitment for their street trees, ensuring that the mature, attractive veteran and heritage trees in communities will be replaced once they have outlived their natural lives decades into the future. Similarly, many municipalities require that parks department staff be certified arborists who can maintain the health of street trees.

This involves pest management, as in the case of controlling the cherrybark Tortix moth that infects ornamental cherry trees (*Prunus sp.*), that have been commonly planted in cities. (It also attacks other fruit trees.) These ornamental trees have showy blossoms in spring but do not produce fruit that could be hazardous to traffic if it falls onto streets. A control is to use pheromones, chemical communicators used to attract the opposite sex. The female scent is placed onto a small cardboard tent suspended on a branch. The male is attracted to the tent by the smell and is caught on a sticky substance [illegible] on the inside walls.

Other insect pests that crawl up from the ground, such as [illegible] worms, inchworms, gypsy moth and tussock moth, can be controlled by using a pest barrier of tanglefoot (a mixture of castor oil, waxes and resins) that is applied in a ring around the trunk of a tree and traps the crawling insects as they climb the trunk to feed on leaves, buds and fruit. A conspicuous band of white cloth is used to form a tight seal with the bark and is held in place with a plastic wrap painted with tanglefoot that the insects are forced to crawl onto on their way up the trunk.

A pheromone trap for the cherry Tortix moth. The trap consists of a milk carton cut into the form of a tent and scented with the smell of the female. It is also coated with a sticky substance that traps the male moth when he comes looking for the female.

Roads and Infrastructure

Municipalities can take care to avoid areas of importance for biodiversity when locating roads or other services through the city.

Greenways

Linking green spaces in the city is an important component of many OCPs for recreational purposes. Greenways also help conserve urban biodiversity through connectivity. Municipalities may require that a development provide for dedicated walkways as part of the approval process.

Greenways management involves maintaining healthy ecosystems, risk management and public enjoyment (Stewardship Series 1995):

Maintaining healthy ecosystems: This includes limiting human disturbance by restricting access in places, putting up barriers and signage

Signage promotes the value of a green space and a map improves safety.

for education. Vegetation management can limit pesticide and herbicide use, reducing cutting and limiting the spread of exotics. Stormwater management of greenways can include limiting impervious surface in the vicinity and implementing watershed scale planning. Water quality management includes plantings of riparian vegetation, sediment traps, erosion protection and vegetated swales for infiltration.

Risk management: Water facilities, such as ponds, should have shallow edges and signage. Recreation trails can be made safer by making available trail guide maps, signage and handrails. Natural hazards can be well marked and alternative routes provided. Use conflicts such as nature walks versus roller-blading and cycling can be resolved through public consultation. There can be a voluntary assumption of risk on the part of the user through signage that limits liability.

Enjoyment: User benefits can be maximized by involving the public, especially nongovernmental organizations, in restoration and maintenance. Programming and special events could encourage greater use.

Other Municipal Tools

In addition to the zoning bylaws to regulate land use, building restrictions and other measures described above, municipalities and regional governments can also conserve biodiversity through the:

- control of water reservoirs
- treatment of waste water
- maintenance of sewers and storm drains
- use of integrated pest management
- application of watershed planning
- use of integrated stormwater management planning
- fostering of best management practices programs

City of Santa Monica — A Sustainable Community

In 1994, the city of Santa Monica, CA launched its Sustainable City Program (www.ci.santa-monica.ca.us/environment/policy/). It wanted to ensure that it could continue to meet its needs, environmental, economic, and social, without compromising the ability of future generations to do the same. The program was designed to help the community to think, plan, and act more sustainably. The program is composed of 9 guiding principles and 8 goal areas. Within each of the goal areas are specific targets with deadlines attached.

The guiding principles are:

- The Concept of Sustainability Guides City Policy
- Protection, Preservation, and Restoration of the Natural Environment is a High Priority of the City
- Environmental Quality, Economic Health and Social Equity are Mutually Dependent
- All Decisions Have Implications for the Long-term Sustainability of Santa Monica
- Community Awareness, Responsibility, Participation and Education are Key Elements of a Sustainable Community
- Santa Monica Recognizes Its Linkage with the Regional, National, and Global Community
- Those Sustainability Issues Most Important to the Community Will be Addressed First, and the Most Cost-Effective Programs and Policies Will be Selected
- The City is Committed to Procurement Decisions that Minimize Negative Environmental and Social Impacts
- Cross-sector Partnerships Are Necessary to Achieve Sustainable Goals

The goal areas are listed below with some of the specific targets:

Resource Conservation

- 70% of year 2000 landfills to be diverted (recycled, re-used, composted, etc) by 2010.
- 20% reduction in city-wide water use by 2010.
- 25% of energy should come from a renewable source by 2010.
- 100% of all building should be LEED™ certified (an environmental or efficiency certification) by 2010.

Environmental and Public Health

- Reduce wastewater flows 15% below 2000 levels by 2010.
- By 2005 all significant emissions sources in the city should be identified.
- Increase permeable surfaces in order to decrease urban run off.
- Increase percentage of locally produced organic produce.
- Increase percentage of local restaurants that buy from local farmers' markets.

Transportation
- An average person per vehicle of 1.5 in all companies greater than 50 employees by 2010
- Percent of total miles of city arterial streets with bike lanes to increase by 35% by 2010.
- 10% reduction in the average vehicles per person by 2010.
- Increase in percentage of residents using bikes, public transit, alternative fueled vehicles, and car-pooling.

Economic Development
- No single section should represent more than 25% of total economic activity.
- Jobs-to-housing ratio should approach 1.
- Job creation should be increasing.

Open Space and Land Use
- Increase in the amount of open space, especially permeable open space.
- Increase in tree canopy cover.
- Increase in percentage of parks within .25–.5 miles (0.4–0.8 km).
- Replace turf in public lands with regionally appropriate plants.

Housing
- Increase in affordable housing to all income levels.
- Increase in new housing within .25 miles (0.4 km) of open space, transit, and grocery stores.
- Increase percentage of green housing.

Community Education and Civic Participation
- Increase voter participation to 50% by 2010.
- Increase in percentage of residents volunteering.
- Increase in percentage of residents participating in community events.
- 25% of all residents understand the ecological footprint of Santa Monica and their contribution towards it.

Human Dignity
- Decrease in crime rate per capita.
- Decrease in the percentage of people that need to work over 40 hours per week to meet basic needs.
- Increase percentage of residents with healthcare insurance.
- Increase in percentage of homeless who are served by city shelters.

Many of the above goals lacking specific deadlines are to be decided upon this year (2003).

Santa Monica is one city that has realized the impact human beings are making on the planet and has taken concrete steps to do something about it. The environment is in the forethought of every decision that is made. They are leading by example, and providing tools so that others can follow.

Provincial and State

Perhaps one of the most powerful tools available to provincial and state governments to conserve and protect urban biodiversity is to provide the enabling legislation for the municipalities to enact their bylaws. Provincial and state governments also have a number of tools for them to take a direct role. In British Columbia, Canada, for example, there are a number of acts that are relevant, such as:

1. Waste Management Act
   - cannot introduce waste that causes pollution (alter or impair usefulness of environment)
   - need a permit to deposit or discharge waste
2. Environment Management Act
   - enables province to order an environmental assessment
   - provides authority for the Environmental Appeal Board
   - allows Minister to declare an environmental emergency
3. Water Act
   - province authorizes the use and diversion of water resources
4. Environment Protection Act
   - goal of zero discharge and polluter-pay principle
   - encourage life cycle management of products
   - toxin-use-reduction legislation and clean-up of contaminated sites
   - ambient air, water and soil environmental quality criteria
   - market-based incentives (taxes, pollution credits, "seals of approval")
   - whistle blowers protection
5. Environmental Assessment Act
   - regulatory framework for examining a variety of hydro, thermal energy, and mining projects
6. Fish Protection Act
   - protection of fish of no commercial value
     - includes non fish bearing nutrient streams

Federal

Canada

Environment Canada

Environment Canada has four priorities: biodiversity, climate change, protection of the ozone layer, and toxic substances. However, Environment Canada is more active in the latter three priorities. It does not have any responsibility when it comes to making decisions on conserving biodiversity. Its main role is to provide science based recommendations and leadership to produce unified national objectives (www.biodiv.org, www.climatechange.gc.ca, www.nrcan-rncan.gc.ca, www.oag-bvg.gc.ca/domino/cesd_cedd.nsf/html/menu6_e.html). Below are some examples of actions they are taking in regard to the other priorities.

- Canadian Environmental Protection Act (CEPA), the goal of which is to contribute to sustainable development through pollution prevention and to protect the environment and human life from risks presented by toxic substances.

Canada ratified the Kyoto Protocol at the end of 2002. Canada has committed to reducing its greenhouse gas emissions to 6% less than 1990 levels by 2012 (http://www.iisd.ca/climate/)

- Some of the ways the government of Canada aims to achieve this goal that are relevant to the urban environment are:
  - Reduce emissions from urban transportation by reduction of cars, shift to less greenhouse-gas-intensive travel alternatives.
  - Encourage more energy efficient buildings, both commercial and residential
  - Improve standards for energy efficiency in appliances
  - Encourage development of alternative fuels technologies such as fuel cells and green power.
- Ozone depleting substance regulations were introduced in 1998
- EcoAction Funding Program. This arose out of the Agenda 21 program. It provides funding to non-profit groups for projects in three areas (air quality and climate change, water quality, and natural environment) that will provide positive, measurable impacts on the environment.

Commissioner of the Environment and Sustainable Development

The Commissioner is responsible for making sure that the Government of Canada is meeting its environmental commitments and greening its policies. To date the Commissioner has not given the government a very good review. In 1998, biodiversity initiatives were reviewed and found mostly lacking. In 2000, a follow up of this was completed. There is no federal implementation plan for biodiversity. Many of the federal modules in place have not adequately addressed important issues such as resource allocation, time frame, and performance indicators. There is still no "ministerial home" for biodiversity which, in some sense, means that no one is responsible or accountable.

Natural Resources Canada

Responsible for fostering the integrated management and sustainable development of Canada's natural resources.

Parks Canada

The mandate of Parks Canada does not use the word "biodiversity", however the Canada Parks Act 2000 states that "maintenance or restoration of ecological integrity, through the protection of natural resources and natural processes, shall be the first priority of the Minister when considering all aspects of the management of parks."

No flora or fauna may be removed from a park.

Agriculture Canada

Agriculture Canada has committed to work with industry to improve environmental performance in four specific areas: water, soil, air and biodiversity.

Urban areas often figure significantly in the survival of some stages of the life histories of populations or entire species such as this Adams River sockeye salmon that depends in part on habitat in Greatever Vancouver, British Columbia.

Additional Federal Legislation:

1. National Standards: Constitutional Authority over the Environment
   - Federal government has the authority to make laws for the "Peace, Order, and Good Government" of Canada (POGG)
2. Canadian Environmental Assessment Act (CEAA)
   - Requires that environmental impact assessments be required for all large-scale Federal projects; this is especially significant in an urban context when applied to airport expansions
3. Canadian Environmental Protection Act (CEPA)
   - governs activities in federal jurisdictions; e.g., cross-border air pollution, ocean dumping, regulation of toxic chemicals
4. Fisheries Act
   - an offence to alter, disrupt, or destroy fish habitat
   - pollution extends to all waters frequented by fish
5. Species At Risk Act (SARA)
   - protects habitat of endangered species
6. Additional Federal Legislation
   - Transportation of Dangerous Goods Act
   - Pesticides Control Product Act
   - Migratory Birds Convention Act
   - Hazardous Products Act

Green Roof Tops

The Canadian Mortgage and Housing Corporation, Environment Canada, and National Research Council have been doing research into the benefits of roof top and vertical gardening in Canada. In the past this practice has been used more widely in Europe than in North America, with 10 million square metres of green roofs developed in Germany alone in 1996. All sources agree that green roof tops are an excellent way to mitigate many of the environmental problems we have in urban areas:

- www.cmhc-schl.gc.ca, www.cityfarmer.org/rooftop59.html
- www.durable.gc.ca/communication/brochure/climatechange1_e.phtml
- www.climatechangeconnection.org/pages/subpages/urban_releaf.html irc.nrc-cnrc.gc.ca/newsletter/v7no1/rooftop_e.html
- www.greenroofs.ca.

The benefits associated with these types of gardens are extensive:

- Improve air quality and reduce CO_2 emissions
- Delay storm water runoff, which can reduce the amounts of sewage overflows and flash floods, and decrease demand on groundwater resources
- Increase habitat for birds and insects, which can lead to an increase in biodiversity in urban areas
- Insulate buildings, reducing energy use associated with heating and cooling and also reducing the effects of the "Urban Heat Island"
- Increase private outdoor green space, which increases quality of life
- Block movement of dust and trap airborne particles
- Protect roof membranes from UV light, which extends their lifetime and reduces costs
- Increase aesthetic appeal and increase property values
- Provide sound insulation

Small trees, shrubs and perennials can transform rooftops into productive habitat.

European studies maintain that rooftop gardens retain 70–100% of the precipitation during the summer and 35–50% during the winter. These gardens store precipitation until used by the plants or returned to the atmosphere through evapotranspiration. This also helps to reduce the urban heat island effect by 3–4 °C, since the vegetation can reduce the amount of re-radiated heat by up to 90% on a hot summer day.

The preliminary findings of the CMHC study show that green roofs reduce the amount of heat entering a building during the summer by 85% and the amount of heat leaving the building by 70%. The daily temperature fluctuations on the roof were reduced from 46 °C to only 6 °C, which leads to a longer life of the roof membrane.

There are several places in Canada that already have green roof tops.

- The Vancouver Public Library

- Ryerson Polytechnic, Toronto
- A parking garage in Quebec City
- An apartment building in Toronto where residents grow food
- Toronto City Hall (demonstration and research site for green roofs)
- Oak Hammock Marsh Interpretive Centre, Manitoba

The Toronto demonstration project consists of two sites; one at Toronto City Hall and one at the Eastview Neighbourhood Community Centre. This is a $1 million, three-year project that proposes to overcome technical, financial, and information barriers and currently prevent the widespread use of green roofs.

There are eight demonstration plots on the roof of Toronto City Hall amounting to 3,200 ft^2 (303 m^2). The plots are divided up as follows:

- Black oak savannah, a native prairie ecosystem that is now endangered in North America. These plants are drawn from the local High Park gene pool
- Two extensive plots featuring a variety of *Sedum* and alpine perennials, typical of industrial and commercial applications
- Two semi-intensive plots featuring a variety of flowering plants and shrubs
- Native butterfly and bird habitat
- Two urban agricultural plots featuring perennials and annuals

The Eastview Neighbourhood Community Centre planting is a 5,000 ft^2 (474 m^2) extensive green roof. Extensive green roofs are characterized by having minimal growing medium depth, weight, plant diversity and maintenance.

The project will be monitoring for improvements in air quality, energy efficiency from reduced heating and cooling requirements, reduction in ambient temperatures (how does this affect the urban heat island), storm water retention, and plant survival.

United States

The Environmental Protection Agency

The Environmental Protection Agency's mission is to protect human health and safeguard the natural environment (land, air, and water). They provide leadership in research, education, and assessment. They work closely with other agencies (federal, state, and First Nations groups) to enforce existing environmental laws and to develop new ones. (www.epa.gov)

The EPA is guided by several important pieces of legislation. A few are listed below:

- The Clean Air Act
- The Clean Water Act
- The Endangered Species Act
- The Oil Pollution Act
- The Pollution Prevention Act
- The Safe Drinking Water Act

The EPA has developed several programs to address these environmental concerns. However, many of the programs serve only to provide information and many others are voluntary.

The Energy Star Program — is a program to help businesses and individuals protect the environment through energy efficiency. By choosing appliances with the energy star rating, people can save about one third on their energy bill while reducing greenhouse gas emissions.

Climate Leaders Partnership — is a voluntary program that encourages companies to make long-term comprehensive climate change strategies.

Green Power Partnership Ltd — is a voluntary program to encourage the reduction in emissions by promoting electricity generation from renewable sources.

Pesticide Environmental Stewardship Program — is a voluntary program that forms partnerships with pesticide users to reduce risks (health and environmental) and to implement pollution prevention strategies.

The Nature Conservancy

The Nature Conservancy aims to preserve plants, animals, and natural communities that represent the diversity of life by protecting the land and water they need to survive. Since 1951 they have protected 98 million acres around the world. (http://nature.org/) The Nature Conservancy has several main priorities:

- **Climate Change Initiative** is working to reduce the effects of climate change on natural areas by protecting and restoring forests and developing adaptation strategies for ecological systems.
- **Fire Initiative** is working towards restoring fire-altered ecosystems.
- **Fresh Water Initiative** is working to combat the two major threats to the freshwater supplies: ecologically incompatible water management and unsustainable agricultural practices.
- **Invasive Species Initiative** aims to control the threat to biodiversity by invasive, non-native species using a combination of prevention, eradication, restoration, research, and education.
- **Marine Initiative** tries to protect the animal and plant life of the marine environment for both ecological and economic benefit.
- Other initiatives including the Migratory Bird Program and the Ecotourism Program.

National Parks Service

The NPS is the manager of 79 million acres of parkland. Its primary mission is to limit human influences in these areas so that natural ecosystems are controlled by natural processes. Natural parks are samples of natural diversity and should be maintained as insurance against the continued loss of biodiversity due to human actions worldwide. (www.nps.gov)

United States Department of Agriculture

The USDA is the steward of 192 million acres of national forests and range land in the US. It is an umbrella for many federal agencies dealing with forests, agricultural land, nutrition, and food safety, to name just a few. (www.usda.gov)

The USDA Forest Service's mission is to sustain the health, diversity, and productivity of the forests and grasslands in order to meet current and future needs. They use an ecosystem approach to management that includes ecological, economic, and social factors. Maintaining species diversity is a priority in the forest and rangeland and in cities. They are a leader in research of the positive air quality and energy saving effects plants can have on urban environments. (www.fs.usda.gov). The Natural Resources Conservation Service is also under the umbrella of the USDA.

Agricultural Research Service has developed many programs to try to understand the effects of agriculture on the environment and to minimize them. These include topics such as air quality, global change, water quality and management, soil resource management, pesticides, and food safety.

Regular monitoring for turbidity, dissolved oxygen and nutrient levels is essential to maintaining water quality.

International

Agenda 21

The Earth Summit in Rio de Janeiro in 1992 was the largest meeting of world leaders ever to focus on environmental issues. These leaders agreed upon a strategy for sustainable development, one that would meet our needs while ensuring that there would be a healthy and viable Earth for future generations. Two binding agreements were signed, the Convention on Biological Diversity and the Convention on Climate Change.

The Convention on Biological Diversity establishes three main goals: the conservation of biological diversity, the sustainable use of its components, and the fair and equitable sharing of the benefits from the use of genetic resources. This convention recognized for the first time that biodiversity was a "common concern of humankind." The Convention is legally binding and has been ratified by more than 175 countries since 1992. The Convention calls for an "ecosystem approach to the conservation and sustainable use of biodiversity" in which all goods and services provided by nature are taken into account.

Out of the Earth Summit came Agenda 21, which is a plan for action, objectives, activities and means of implementation. There are many objectives of Agenda 21 that are of particular interest when considering biodiversity of urban areas (www.un.org). A few are listed below.

- Chapter 9, Protection of the Atmosphere. There is a need to control greenhouse gases, decrease the use of fossil fuel energies, and promote energy efficient strategies. In terms of transportation, we must promote and develop cost-effective, less-polluting, safer transport systems with special reference to urban transport systems. With respect to the terrestrial environment it is important to recognize that reduced biodiversity may reduce the resilience of ecosystems to climatic variation and air pollution damage. Initiatives are aimed at reducing atmospheric pollution and limiting greenhouse gases, conservation, sustainable management, and enhancement of all sinks and reservoirs of greenhouse gases.
- Chapter 15, Conservation of Biological Diversity, states that even small areas such as gardens are of great importance and policies must be developed to encourage conservation of private lands.
- Chapter 17, Protection of Oceans and Coastal Areas, calls for integrated coastal management. Over 60% of the populations live within 60 km of the shoreline. This calls for improvement of coastal human settlements especially with regard to treatment of sewage and effluent from housing and industry. It also calls for the conservation and restoration of critical habitats.
- Chapter 18, Protection of Freshwater Resources, calls for implementation of urban storm runoff and drainage programs, promotion of recycling and reuse, protection of watersheds, and adoption of city-wide approaches to management of water.
- Chapter 28, Local Authority's Initiatives. Agenda 21 recognized that most of the changes would come not at an international or national level, but at a local level. One of the objectives, therefore, was that all local authorities should have a "local Agenda 21" plan by 1996. As this level of government is

closest to the population, this objective was seen as a way to enter into dialogue with the community, and as on opportunity to educate and encourage the average citizen to make a difference, and to find strategies that would work and be respected.

Convention on Climate Change

Climate change is thought to be the greatest threat to the sustainability of the world's environment. Most of the current climate change is occurring due to human activities. The increases in greenhouse gases (ghg), such as carbon dioxide, are mostly due to emissions caused by the use of fossil fuels; electricity generation, transportation, and agriculture are the culprits. The United Nations Framework Convention on Climate Change (UNFCCC) set out to create a framework to reduce greenhouse gas emissions. It was agreed upon in 1992 at the Rio Earth Summit. In 1997 a protocol was adopted in Kyoto that committed industrialized countries to achieve quantified targets.

Other International Agreements

1. UN Commission on Sustainable Development
   - monitor progress in implementing Agenda 21
2. Canada-US Free Trade Agreement (FTA)
   - concerns: e.g., 20% more licensed active pesticide ingredients in US, and 7 times more pesticide products
3. General Agreement on Tariffs and Trade (GATT)
   - In trying to protect the environment, GATT is violated when:
   - GATT decides that the true intention of the regulation is to restrict trade, not environmental protection
   - the restrictions imposed are more than necessary to protect the environment
   - a country sets standards with no scientific basis
4. North American Free Trade Agreement (NAFTA)
   - concerns: erosion of standards and conservation measures, trade experts (not elected) will influence legislation
5. Montreal Protocol
   - reduce ozone depleting substances
6. International Joint Commission 1909
   - bilateral (US-Canada), authority to act on transboundary pollution - mainly water issues, some air

Nongovernmental Organizations

There are a number of nongovernmental or not-for-profit organizations that work to conserve and enhance natural areas. Some may have an urban focus but many do not, although they support city plantings as well as preserving more natural ecosystems. They include nature clubs and federations, such as the Canadian Nature Federation and National Audubon Society. Another example is the Tree Canada Foundation.

Tree Canada Foundation

Tree Canada Foundation is a not-for-profit organization that provides education, resources, and financial support to encourage Canadians to plant trees in urban and rural areas to help reduce the effect of carbon dioxide emissions.

Tree Canada Foundation planted 1.5 million trees in celebration of the millennium year.

Further Initiatives that Increase Urban Biodiversity

Pesticide Ban

Pesticide is a broad classification for a chemical that is used to kill things humans consider pests: insecticides (to kill insects), herbicides (to kill plants), fungicides (to kill funguses), and rodenticides (to kill rodents) (www.spec.bc.ca, www.region.halifax.ns.ca/environment/index.html, www.sierraclub.ca/national/pest/pesticid.html). Most pesticides are non-specific, killing off the good with the bad, which can greatly reduce the biodiversity of an area. Pesticides are known to cause drastic problems in nature. DDT is an insecticide that was used widely in North America until the early 1970s. It is highly toxic to fish, and contributed greatly to the decline of the Bald Eagle populations. The eagle was on the brink of extinction partly because DDT softened egg shells which then led to death of the offspring.

There is evidence that pesticides are also quite harmful to humans, and especially to children. A National Cancer Institute study in the US indicates that children are as much as six times more likely to get childhood leukemia when pesticides are used in the home and garden (Sierra Club). Children, fetuses, and pets are especially vulnerable to pesticides because they have not yet developed the ability to metabolize and eliminate toxic substances. Children and pets are also closer to the ground and often eat dust, grass, and flowers (which essentially means they are eating pesticides).

Westmount, Que was among the first cities in Canada to introduce a bylaw restricting the use of pesticides on public and private land. The city started with small restrictions in 1994, and by 1999 had a city-wide ban on pesticide use during the summer months.

Halifax Regional Municipality was another leader in this field. In April 2000 they passed a bylaw that would be fully phased in after three years. The by-law calls for a ban on all pesticide use on municipal and residential property. The by-law defines pesticide use as “the application and use of pesticides for the maintenance of outdoor trees, shrubs, flowers, other ornamental plants and turf on the part of a property used for residential purposes or on a property of the municipality.”

The Halifax Regional Municipality provides many resources on ways to garden without pesiticides including a website with alternatives to chemical pesticides, a newsletter entitled "Naturally Green", workshops on how to maintain a sustainable yard, and links to other online and printed resources.

In July, 2002, Port Moody, BC became the first municipality in western Canada to ban cosmetic pesticide use on residential property. Port Moody already has an integrated pest management program for its public lands. For the past 10 years, except in rare circumstances, no pesticides have been used on parks, school yards, playing fields, boulevards or medians. Instead, staff members have learned to plant pest-resistant native species, to use biological controls such as ladybugs against aphids and to plant pest-repelling plants next to vulnerable ones (Society Promoting Environmental Conservation website, www.spec.bc.ca).

These bans will lead to increasing the biodiversity of city. They also provide a great way for people to learn more about sustainable gardening practices in general. All of these municipalities have included a long time interval for education of the public and have provided this education in terms of workshops, websites, and printed materials. The commitment to treating public green spaces in this manner is also in commendable.

17

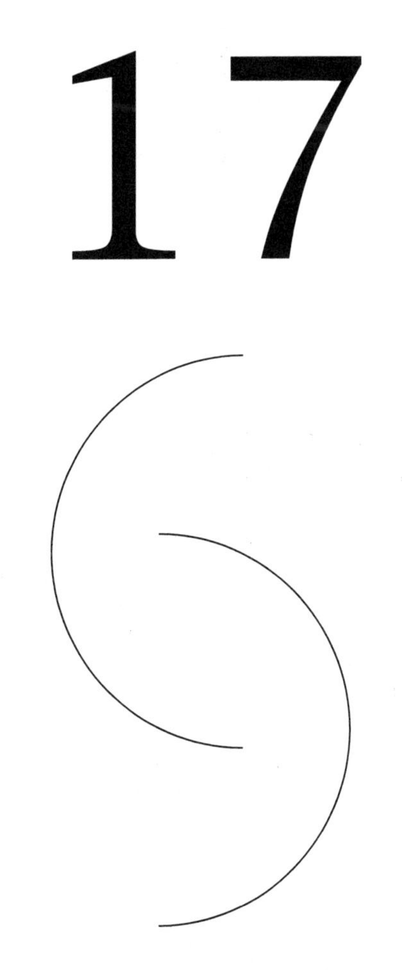

Case Studies

Biodiversity has caught the attention of many countries. In Canada, there have been a series of recent events addressing biodiversity — the Special Places 2000 program in Alberta that was intended to identify areas worth protecting but ended up mired in controversy over accusations of political manipulation; the Lands for Life planning process in Ontario that led to a plan to expand protected areas in the province; and, off the coast of Nova Scotia, the federal government acted to protect a rich area of biodiversity on the continental shelf known as The Gully (Bocking 2000).

In a more urban context, habitat restoration projects are being actively undertaken all over North America and are having beneficial effects on ecosystem structure and function. In general, the habitat restoration projects explored below contribute many things to the ecosystems services discussed in previous chapters. Depending on the type of ecosystem restored and the extent of the project undertaken, the benefits to the region can be very extensive.

An important principle in urban habitat restoration is that not all habitats should be restored. This park is clearly seen as a recreational facility for the community and is not a good candidate for restoration to maximize biodiversity in the region.

These projects promote connectivity. Ecological connectivity is a key component of the case studies that will be presented in this chapter. They not only connect wildlife and plants; community members become connected through a common goal and gain a significant connection to the environment.

Presented here are several case studies that illustrate only a fraction of the habitat restoration and stewardship activities that have taken place and are under way across North America. These projects have all engaged local communities, have lasting benefits to local communities, contribute at varying degrees to nature's services, and improve local ecology and biodiversity.

It will be explored in more detail as an example of a strategy to conserve biodiversity.

Canada

British Columbia

Greater Vancouver Biodiversity Conservation Strategy

Greater Vancouver is rich in biodiversity. It has a marine habitat (Strait of Georgia, Burrard Inlet, Boundary Bay), major watersheds (Fraser River, Brunette River Drainage Basin, and the North Shore watersheds that provide drinking water — Capilano, Seymour and Coquitlam), a major delta and estuary, and diverse upland habitat. Urban growth in the region is confined by the Coastal and Cascade mountains, putting urban development in direct conflict with lands rich in biodiversity.

There had been much discussion and consultation amongst various groups to resolve conflicts as they affected particular issues. For example, the destruction of foreshore habitat along the Fraser River and its numerous tributary creeks resulted in the formation of the Fraser River Estuary Management Plan that primarily conserved salmon habitat through a comprehensive program of research, development guidelines and community action. Similarly, the Burrard Inlet Environmental Action Program applied a similar approach to land use conflicts in Burrard Inlet.

However, in 1999 there was a move to conserve biodiversity as a whole in the Greater Vancouver landscape. At that time a partnership to develop a biodiversity conservation strategy was formed between groups including the Georgia Basin Ecosystem Initiative, the BC Ministry of Water, Land and Air Protection and Environment Canada's Canadian Wildlife Service. Some key issues to be addressed were habitat loss, habitat fragmentation, and disturbance by invasive alien species.

Initially, the group addressed some basic questions:

1. Who is involved — who will lead, what is the target audience, who is going to implement recommendations
2. What is the planning scale — municipal, regional, ecological management boundaries that transcend political boundaries
3. What will be conserved and why — indicator species, endangered species, ecosystems

4. Biophysical inventory needs — existing information, gap analysis, remote sensing,
5. Integration of socio-economic factors — managing growth, transportation, agriculture

It was also important to establish how the strategy would be positioned relative to all of the other priorities and planning initiatives in the region. For example, it was understood that the strategy needed to:

- Satisfy the mandates of existing and potential stakeholders such as agencies from various levels of government, nongovernmental organizations and industry
- Use spatial and planning methods that are standardized and recognized as valid scientifically and accepted locally and internationally
- Recognize that moving from vision to reality occurs through local decision making

The development of the GVRD Biodiversity Conservation Strategy evolved into a four phase process. These are (GVRD 2003):

- Phase 1: Background and Scoping: Identifying key issues, objectives and components with an introductory stakeholder workshop.
- Phase 2: Habitat Mapping and Analysis: Compiling a GIS database and catalogue of protected areas, developing a biodiversity map of the Greater Vancouver Region using remote sensing imagery for ecological classification, developing a set of indicator species and habitat requirements for future monitoring
- Phase 3: Developing Strategic Directions for Biodiversity: Examining the current institutional framework and issues involved in conserving biodiversity, preparing a socioeconomic study highlighting benefits of biodiversity conservation, determining priorities and strategic directions to advance the biodiversity conservation strategy
- Phase 4: Developing a Strategy for Biodiversity Conservation: Identifying core habitat areas and corridors, developing tools, defining a monitoring program and developing a public communications and education strategy

Phase 4 will be completed in 2004. Presented here is a brief summary of the recommendations resulting from Phase 3 (from GVRD 2003b):

1. Review of Key Biodiversity Conservation Issues, Roles and Responsibilities.

   A. Administrative Framework
   - Become better informed on biodiversity conservation issues
   - Identify overlaps and provide linkages to address redundancies
   - Foster consistency among related issues (e.g., a toolbox approach)
   - Facilitate and maintain partnerships amongst those involved

   B. Adaptive Management
   - Monitor progress
   - Make options available for continuity of multi-year projects

C. Bringing Biodiversity to the Forefront
- Integrate biodiversity considerations into all other regional functions and strategies
- Link the strategy to other regionally-directed programs and plans
- Increase awareness of biodiversity to the public
- Offer meaningful incentives to land owners and users to encourage biodiversity

2. Socio-Economic Values of Biodiversity in the Greater Vancouver Region.

A. Collect Local Data
- Define use values (e.g., recreation — users per year)
- Identify and quantify biophysical improvements of ecosystem services such as air quality

Identify and quantify psychological, cultural and spiritual values

B. Apply Local Government Resources and Tax Revenues
- Identify and implement fees and taxes to generate revenues that help support biodiversity conservation efforts

C. Encourage Biodiversity Conservation on Private Land
- Provide subsidies to farmers that manage land for wildlife (e.g., food crops for overwintering waterfowl as with the local Greenfields Program
- Develop taxes, user fees and final demand interventions (e.g., green certification) to encourage biodiveristy

3. Priorities, Opportunities and Strategic Directions.

A. Broad Strategic Level Recommendations
- Provide regional coordination (e.g., create a Coordinating Body0
- Address gaps, overlaps and conflicts in the administrative framework
- Build on what's already in place
- Monitor success

B. Priority Level Recommendations
- Identify short-term (1–2 years), medium-term (3–5 years) and long-term (greater than 5 years) priorities
- Maintain biodiversity values in areas already protected
- Create a common vision and strategy that includes an operational definition, realistic goals and objectives, and a framework to coordinate initiatives
- Increase public education and awareness regarding biodiversity concepts and initiatives
- Conserve significant non-secured habitats
- Maintain connectivity and minimize fragmentation
- Maintain air and water quality

C. Implementation Framework
- Clarify biodiversity benchmarks
- Finalize a list of indicators and targets
- Develop a mechanism to report progress

Phase 4, under way in 2004, involves:

1. Developing a comprehensive base map of biophysical resources for the region
2. Examining the Still Creek Watershed as a case study
3. Identifying strategic options, such as core areas, hot spots, linkages and connectivity

A final step will be a thorough investigation of policies and tools available or needed to implement the strategy, and then proceed with its implementation.

The detailed planning process to develop the Biodiversity Conservation Strategy raised many questions that challenge the limits to our knowledge and understanding of urban biodiversity. These include:

The Barn Owl is a threatened species in British Columbia and its breeding habitat can be improved with a nest box.

- What degree of accuracy makes sense biologically when mapping urban habitat? One square metre, one hundred, one hectare? Increased resolution may be possible through remote sensing, but it may be unnecessary. It may in fact be inadvisable because it is more expensive and it may cause confusion.
- What level of biophysical information is necessary to manage a green space? Is a complete, detailed biophysical inventory necessary for a green space that is too small or disturbed to contribute meaningfully to regional biodiversity?
- Are there thresholds beyond which biodiversity thrives, or below which biodiversity plummets? How do we identify green spaces on the threshold where it is important to conserve every last part to prevent a crash in biodiversity, or where by adding a relatively small amount more to the green space we would see a dramatic increase?
- What should be done about rare and endangered species? When is it worthwhile to invest large amounts of human resources and money into conserving one species at the expense of conserving other species and habitat to maximize biodiversity?

- How much local information is necessary and how much can be inferred from theoretical treatments of patch size and connectivity?

Spanish Banks Creek Daylighting, Vancouver

Daylighting, or uncovering buried watercourses, is increasingly becoming a popular activity in urban areas. Where once watercourses were covered over and replaced with storm drains — taking away valuable fish and wildlife habitat — they are now being uncovered to improve habitat and urban ecosystem health. There are many examples of stream daylighting from within the Lower Mainland and across North America, allowing us to realize that some of the negative impacts on fish habitat that are associated with urban development can be reversed or, at the very least, mitigated.

Spanish Banks Creek, in Vancouver's Point Grey neighbourhood, is just one example of salmon returning to spawn after decades of absence because of a very successful daylighting program. In 2001, after nearly 80-years absence, salmon returned to this creek to spawn. The 52 m long daylighting project included removing an old culvert and recreating natural stream complexity. Key additions to the stream and its bank are: gravel to provide spawning material, woody debris to provide cover for young fish, streamside or riparian vegetation to stabilize the banks, and the addition of pools and riffles to add other key habitat components. The Spanish Banks Streamkeepers, who initiated the daylighting project, also released coho and chum salmon fry into the creeks with the hopes of revitalizing salmon populations in the stream. In addition to the salmon returning to the stream to spawn, river otter and mink have also been spotted enjoying the riparian vegetation along the edges of Spanish Banks Creek.

Many local schools and community members have been and continue to be involved with the project. The Department of Fisheries and Oceans has initiated a program where 20 classrooms are rearing chum salmon fry in classroom incubators. These fry are intended to be released into the creek.

The success of the Spanish Banks Creek daylighting project can be attributed to the hard work of the Spanish Banks Streamkeepers, Department of Fisheries and Ocean and other community members Spanish Banks Creek Daylighting. (http://www.gvrd.bc.ca/sustainability/casestudies/spanishbanks.htm. Accessed: November 10, 2003; The Vancouver Courier. Salmon Return to Spanish Bank stream after 80-year absence. http://www.vancourier.com/121201/news/121201nn8.html. Accessed: November 15, 2003; Spanish Banks Creek, Vancouver. http://www.urbanstreams.org/creek_spanishbanks.html. Accessed: November 15, 2003.)

Garry Oak Restoration Project — Greater Victoria

The Garry Oak Restoration Project (GORP) is a public education program that is aimed at educating community members about Garry oak ecosystems as well as habitat restoration. Garry oak ecosystems have been decreasing in size rapidly due to invasion of introduced species, urban development and suppression of natural fire regimes.

GORP uses a "living laboratory" approach to environmental education. Several parcels of Garry oak ecosystems that are owned by the municipality of Saanich have been developed and maintained as demonstration sites.

Objectives of individual demonstration sites range from restoring open Garry oak woodlands to enhancing rocky outcrops with the possibility of potential restoration of the associated Garry oak savannah, to planting native species and removing non-native species.

Demonstration sites are located at McKenzie and Arrow Roads, the Victoria Association of Community Living on Cedar Hill Cross Road at Merriam Road, Playfair Park and Mount Tolmie.

GORP and other Garry oak ecosystem awareness organizations have been successful at educating community members on the ecosystems and their status in Greater Victoria. Recently an educators' booklet designed for teachers, naturalists and community members was completed and distributed to the community.

Common camas (Camassia quamash) and white fawn lily (Erythonium oregonum) are characteristic of the meadow associated with the Garry oak ecosystem.

The Green Links Project — Greater Vancouver

Green Links is a project of the Douglas College Institute of Urban Ecology (Schaefer 1999a, b). It addresses the problem of urban habitat fragmentation by establishing and maintaining ecological corridors in urban areas throughout Greater Vancouver. The goal of the Green Links Project is to maintain the viability of natural ecosystem fragments in urban areas through the conservation biology practice of connectivity and to use a holistic approach that involves the community.

A Green Link is a type of greenway, but it differs from most in that the movement of people is not an essential part of the design, enabling backyard habitats to be incorporated. The fragmentation of urban wildlife habitats is becoming a particular problem for BC's Lower Mainland. Over the past 10 years Greater Vancouver has grown to 1.8 million people, with the population expected to almost double to 3.1 million by 2021. These same estimates show that if current land-use practices are to continue, land for single family housing will run out within 15 years.

The Greater Vancouver Regional District's response to this situation is the "compact metropolitan model," which will concentrate future growth into regional centres in Vancouver, Coquitlam, Burnaby, New Westminster and North Surrey/North Delta. While this plan will have the laudable effect of protecting the Green Zone around Greater Vancouver, it will significantly increase the amount of denser, low-rise housing in these municipalities. If left unchecked, this growth will contribute to the already advanced state of fragmentation in the Lower Mainland's wildlife habitats.

Green Links regards urban areas as a home to both people and nature. It operates on the principles that cities usually develop in areas of important wildlife habitat such as estuaries and floodplains, that habitat fragments within cities contribute significantly to the viability of the greater ecosystems in which they occur, and that environmental stewardship by individuals contributes incrementally to the productivity of the area.

Green Links recognizes that important wildlife habitat is occurring in cities, where the primary land-use may be economic, social and cultural. Thus, both wildlife and human values are considered. Environmental education activities and publications highlight the diversity and value of natural habitat in developed urban landscapes.

Green Links is also a relatively new approach in its holistic strategy, encouraging both concrete action, such as habitat restoration, and a change in values (through writings and art), resulting in more environmentally sustainable lifestyles.

In order to establish links, biodiversity is increased through plantings of native vegetation (primarily shrubs and perennials). The plantings are planned and organized to involve the community at large and organized groups such as service clubs (e.g., Optimists, Rotary), youth clubs (e.g., scouts, guides), municipal and regional governments (e.g., City of Burnaby, Greater Vancouver Regional District) and nongovernmental environmental organizations (e.g., Vancouver Natural History Society, Burns Bog Conservation Society).

The primary objective is to increase the ecological value and biodiversity of urban wildlife habitats and green spaces in the Lower Mainland of British Columbia in general, and in Coquitlam in particular. There are two secondary objectives for Green Links. One is to increase the value of green spaces to the community. A second is to reduce the ongoing maintenance costs (and demand for potable water) associated with managing green spaces and home gardens.

This will be accomplished primarily through planting native vegetation in utility corridors and backyard habitats designated as Green Links corridors. These plantings will create theme habitats, such as butterfly gardens, hummingbird gardens, multiple species habitats and low maintenance habitats.

Two measures of biodiversity are being used. One is the Simpson's Index. The target is to use Green Links to raise the average biodiversity index (Simpson's) for birds (used as an indicator of overall biodiversity) by 30% (from 10 to 13) over a 10-year horizon, or an average of 3% per year in the short term.

A second measure is the presence of indicator species. The assumption is that encouraging such species with more sensitive habitat requirements encourages more numerous species with less sensitive requirements. Examples of such indicator species may be Dark-eyed Junco (*Junco hyemalis*) for ground cover, Rufous-sided Towhee (*Pipilo erythrophthalmus*) for shrub layer, Rufous Hummingbird (*Selasphorus rufus*) for nectar producing flowers, and Yellow Warbler (*Dendroica petechia*) for tree canopy habitat.

The following activities are being used to increase connectivity:

- Restoration of native plant species. Several types of plantings are being used, depending on the conditions and requirements of each specific site. Various planting programs possible are:
  - butterfly and hummingbird gardens (herbaceous, low growing, plants)
  - multiple species habitats (incorporating shrubs, such as native beaked hazelnut for Steller's Jay and squirrels
  - green space maintenance (ground cover and shrubs to out compete nuisance species)
- Removal of invasive species such as Scotch broom
- Constructing multi-use pathways
- Refuse cleanups
- Creating interpretive sites
- Conducting community workshops and erecting bird and bat boxes

Green Links is working in three demonstration sites — two are utility corridors (Coquitlam and Surrey), and the third is through a matrix of residential development (Burnaby). The Coquitlam corridor (Figure 17.1) is the prototype and is the one being reported on here. The Coquitlam corridor is approximately 8 km long and 100m wide and 128 hectares in area. The land is primarily owned by the City. We work with BC Hydro to ensure that plantings meet height and species requirements for the utility. Green Links increases connectivity between five ecosystem fragments:

Colony Farm (65 ha), a habitat of field and marsh adjacent to the Coquitlam River, was recently made into a Greater Vancouver Regional District Park in recognition of its natural value.

Riverview Lands (31 ha), which posses an ecologically unique arboretum stewarded by the Riverview Horticultural Society. The variety of trees it contains includes every tree species known to grow in British Columbia.

Mundy Park (192 ha), a large municipal park containing a remnant forest and small lake with bog habitat. It is on the top of a moraine marking the boundary between the Burrard Inlet and Fraser River watersheds.

Pinnacle Creek ravine (59 ha), part of the Chine Heights escarpment running between Coquitlam and Port Moody.

Scott Creek ravine (8.5 ha), part of the Westwood Plateau and an important urban salmon stream of the Coquitlam River watershed.

Baseline biophysical inventories have been completed for 14 sites along the corridor in previous work. The utility corridor supports 121 species of plants and 51 species of birds. The Simpson's biodiversity index for birds from the 14

Figure 17.1 The Utility Corridor in Coquitlam, British Columbia

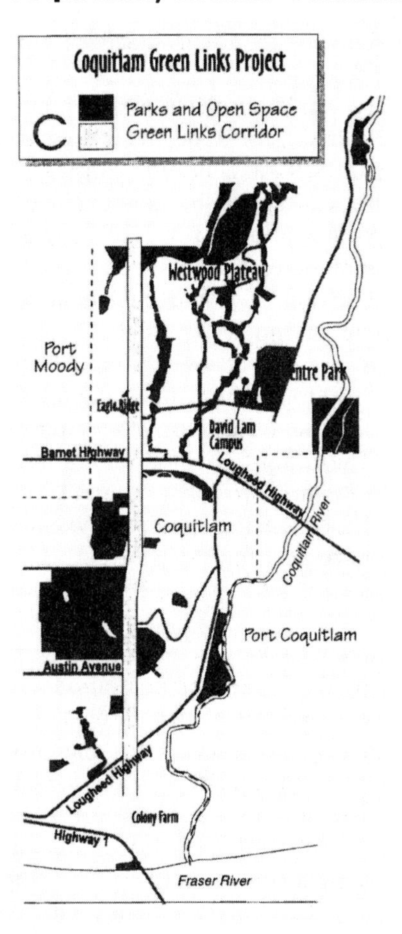

The Utility Corridor in Coquitlam, British Columbia used as the first Green Link Project demonstration site. The light grey area shows the 8 km long utility corridor. The patches of green space it connects are shown in dark grey. Mundy Park is the large green space towards the middle of the corridor by Austin Avenue with Pinnacle Creek Ravine immediately above. (Schaefer and Sulek 1997)

sampling sites along the 8 km corridor ranges from 7.4–16.74, with an average of 10.7. Based on the biodiversity index of 13.0 found at a site second closest to the wilderness fringe of the corridor, a target was set to increase the average biodiversity index for the entire corridor to 13.0 over a 10 year period.

Implementation activities in 1997 in the Coquitlam corridor included the construction of over 1 km of multi-use pathway, water channeling (one location), removal of invasive species (Scotch broom, Himalayan blackberry, purple loosestrife) at three locations, a plant salvage of 500 trees at one location and planting native vegetation at seven locations, with about 3,000 plants covering approximately 6 ha.

A community survey of 2,300 households resulted in 327 respondents, the majority of whom appreciated the green spaces in their community and supported habitat enhancement work.

In 1996/97, the Green Links Project as a whole, encompassing all three corridors, resulted in the planting of about 6,000 plants covering about 10 hectares, presentations to 2,500 school children, construction of 350 bird and bat boxes, community workshops attended by about 250 people, 70 newspaper and magazine articles, a symposium attended by 120 people representing over 30 organizations, 100,000 seeds of perennials mailed to households, over 1,000 plants salvaged and the implementation of a native plant propagation program in four schools. About 600 people attended 12 public speaking engagements, and a Green Links Display was present at over 20 public events.

Baseline biophysical inventories have been completed for 14 sites along the corridor in previous work. The utility corridor supports 121 species of plants and 51 species of birds. The Simpson's biodiversity index for birds from the 14 sampling sites along the 8 km corridor (Figure 17.1) ranges from 7.4–16.74, with an average of 10.7. Based on the biodiversity index of 13.0 found at a site second closest to the wilderness fringe of the corridor, a target was set to increase the average biodiversity index for the entire corridor to 13.0 over a 10 year period.

The initial step in Green Links was to consult with the numerous community groups in the areas of the three demonstration projects. Each was given a presentation on the project and, in return, provided a letter of endorsement. They later served on Advisory Committees for the project and volunteered for planting events. Community involvement was further cultivated through a community survey to gauge the level of support for green space enhancements.

The letters of endorsement were used to obtain project funding from various corporate and government sources. Corporate support came from utility companies — Terasen Gas and BC Hydro, banks with community environmental grants — Canada Trust and VanCity Credit Union, gardening supply shops, and others. Government support was in the form of youth employment programs.

The funding supported the human resources to conduct biophysical inventories and to organize and conduct community plantings. Plant stock was either purchased from wholesale nurseries or salvaged from development sites. Only native plants were used, primarily low growing perennials and shrubs because of height concerns in electrical utility corridors. Some site preparation and much of the trail construction was done by the City of Coquitlam.

A nursery was established at Colony Farm to store salvaged and nursery plant stock, to raise immature stock and cuttings. Bird and bat boxes were constructed by community members in workshops where they were provided with kits by Green Links. In addition to planting, land acquisition, restrictive covenants and other opportunistic tools can be pursued in the future to secure nodes of habitat to anchor the links. Changes in community values were addressed through public speaking engagements, writing articles for newsletters and magazines, media relations and producing educational material.

Alberta

Carsharing Co-op of Edmonton

In urban centres across Canada, car-sharing programs, or co-ops, are becoming more and more popular. Car sharing is a relatively new phenomenon that allows people who sometimes require the use of vehicle, but don't want the burden associated with owning one, to buy into a car timeshare, or co-operative. Members of car co-ops are able to share the costs associated with running a vehicle, while still being able to enjoy the benefits.

With the help of funding from a variety of sources, including Alberta Ecotrust and Environment Canada's EcoAction program, Edmonton launched a car sharing program in June 2000. Since its inception, this program has been a success. In Edmonton, as in the other communities where carsharing programs operate, the program is located in three high density neighbourhoods: Riverdale, Strathcona/University and Oliver. To join the co-op, potential members are required to pay a deposit that is refundable once the member decides they no longer want to be part of the program. Vehicles are booked in advance by members, and for each vehicle use the member pays a per-hour, -day or -week rate (depending on the length of use) and a per-kilometre rate. One of the financial benefits of the Carsharing Co-op is that members only pay for the vehicles when in use, greatly reducing vehicle costs (EcoAction Success http://www.ec.gc.ca/ecoaction/success_display_stories_e.cfm?story_ID=11181. Accessed: November, 2003; Carsharing Co-op of Edmonton. http://www.web.net/~cce/. Accessed: November 15, 2003; CarSharing Network. http://www.carsharing.net/where.html. Accessed: November 15, 2003).

The benefits of the Carsharing Co-op program are not only financial. The environmental benefit associated with sharing vehicles is significant. As carsharing programs increase in popularity, the number of cars on the road in urban areas, as well as the vehicle use, will decrease. The general public contribute to greenhouse gas emissions primarily through vehicle use and, by encouraging changes in use and providing alternatives, the Carsharing Co-op and other similar programs are beginning to change the number of vehicles on the road and make an impact on greenhouse gas emissions. In addition to reducing greenhouse gas emissions, carsharing programs also (Carsharing Co-op of Edmonton's website (http://www.web.net/~cce/why.html):

- Reduce emissions of sulphur dioxide, nitrous oxides, particulate matter (PM_{10}), volatile organic compounds (VOCs), benzene, toluene, and xylene which all pose health and environmental threats.
- Reduce the financial and psychological stress of owning and operating a vehicle.
- Improve local air quality and health.
- Reduce congestion, sprawl, and wear and tear on the transportation infrastructure.
- Improve urban quality of life.
- Increase pedestrian safety in our communities.

Other Canadian cities that are homes to carsharing programs are Calgary, Halifax, Kingston, Kitchener/Waterloo, Montreal, Quebec City, Toronto, Vancouver and Victoria.

Bow River Valley, Calgary

The Bow River Valley is a unique freshwater ecosystem located in the heart of Calgary. It offers a wide range of recreational activities, and habitat for fish and other wildlife. This river system is subject to pollution sewage and urban runoff, and this pollution is only expected to increase as the city of Calgary grows. In order to reduce the runoff and its associated pollution, the Elbow Valley Constructed Wetland was created in 1996. The site is approximately 20,000 m^2, and the wetland surface area is 5,000 m^2, with a storage capacity of 2,300 m^2. This wetland allows the natural system to clean stormwater runoff prior to it, flowing into the Bow River. The construction of the wetland began with the creation of stormwater management ponds that are now being used to test natural water purification methods (Johnson 2002).

The most important outcome of this wetland is its educational benefits. School groups and the general public were involved in creating the wetland and it is now used as an outdoor classroom, teaching people about the need for stormwater treatment and the plants and animals that are associated with wetland habitats.

The wetland is also a field research site for the University of Calgary. Research activities include measuring the survival of plants and their growth pattern and rates, and the effectiveness of water treatment and pollutants removed by wetland plants.

Over 40 school groups per year visit the Elbow Valley Constructed Wetland and, through the construction of boardwalks and interpretative signs, community members have the opportunity to visit and learn about the wetland.

Manitoba

Bishop Grandin Greenway — Winnipeg

The Bishop Grandin Greenway (BGG), located in Winnipeg, is a 4.5 km long by 300 m wide corridor that runs adjacent to a busy four-lane road, the Bishop Grandin Boulevard. This corridor is located in the southern part of Winnipeg, and serves as a right-of-way for electrical transmission and as an underground aqueduct. In May of 1999, the concept of this Greenway was formed during a community meeting. Further community interest was then assessed during an open house in the fall of 1999 (Johnson 2002).

The vision for this project is to use the green space on the north and south side of Bishop Grandin Boulevard to develop a pathway that would link communities and connect plants and animals. Community involvement in the planning and implementation of the BGG is a key component in the process of developing the greenway. Many community members would like to see bike paths, walking paths, skateboarding areas and the restoration of natural habitat.

The long-term vision is to see this pathway as part of a larger system of links in the Winnipeg area, including Fort Whyte and Assiniboine Forest.

After workshops and planning meetings, the first planting for the BGG took place in October 2001. In the following month, a public meeting was held to present plans for future restoration. The BGG has, and will continue to involve public input. Over 160 volunteers have participated to date, and hundreds more are expected to take part as restoration along the BGG continues.

Ontario

North Toronto Green Community, Toronto

Green communities are groups of people working to improve the local environment. Examples of green community activities are:

- Conducting in-home visits/consultations to allow people to make informed, cost-effective, environmental choices,
- Installing energy and water saving devices at homes,
- Green space naturalizations or community plantings,
- Selling rain barrels, composters and native plants,
- Distributing educational materials to community members, and
- Providing workshops on various environmental and sustainability topics.

The North Toronto Green Community is a group of community members working together to promote conservation and improve the environment in north Toronto for their children and future generations. This program had small beginnings in 1994, and has grown to have implemented several successful projects within the community (North Toronto Green Community. http://www.ntgc.ca/index.html. Accessed: November 18, 2003).

Once a month, the North Toronto Green Community hosts speakers on various topics that are of interest to members and community members. In addition to the speaker series, this organization has been instrumental in creating the Eglington Park Heritage Community Garden. They provide educational walks, with a focus on the Don River Watershed and its restoration (Lost River Walks), and they provide tours of green gardens. The North Toronto Green Community has been instrumental in implementing many other programs and is working closely with other organizations to help make north Toronto a more greener and more sustainable place to live.

City Hall, Green Roof Infrastructure Demonstration Project, Toronto

Green roofs are a relatively new technique, practiced in office and other buildings in cities across North America, to provide many ecological and economic benefits.

Benefits of Green Roofs

Green roofs provide many benefits to the general public as well as private building owners. The following is a brief list of the benefits of green roofs:

- A variety of economic benefits through job creation for roof installation and cost saving measures,
- Improved air quality,
- Temperature regulation,
- Stormwater retention, water filtration, and reduced runoff,
- Social benefits, including improved aesthetics,
- Preservation of habitat and biodiversity,
- Local food production and
- Several economic, aesthetic and other benefits to individual building owners.

Toronto's City Hall is home to the *Green Roof Infrastructure Demonstration Project*: a $1 million public-private partnership that was launched in 2000. This project is aimed at showcasing the benefits of green roofs in urban areas. The goal of this project is to generate reliable technical data on the benefits of green roof in areas such as energy efficiency, heating and cooling benefits, air quality improvements, stormwater retention, and the role of green roofs in plants survival and maintenance of biodiversity (www.greenroofs.ca).

Components of this green roof include native butterfly and bird habitats, urban and agricultural plots with perennials and annuals, two semi-intensive plots featuring a variety of flowering plants, shrubs and small trees and two extensive plots featuring a variety of *Sedum* and alpine perennials as well as species native to the Black Oak Savannah (a native prairie ecosystem) (Low Impact Development Centre, Inc. http://www.lid-stormwater.net/greenroofs/greenroofs_commercial.htm).

Nova Scotia

Smart Growth Principles for Halifax Regional Municipality, Halifax

As growth in and around urban areas across the country continues, many municipalities and community organizations are exploring smart growth principles as alternatives to typical urban sprawl. As cities across North America grow, the strategies designed to deal with this growth are continually evolving and are often unique to the geographical region.

The Halifax Regional Municipality, like many other urban centres across Canada, is experiencing rapid growth. This fast rate of growth put extra demand on the existing infrastructure and environment.

- Recognizing that new urban and economic growth strategies are necessary, Halifax has embarked upon a major smart growth marketing campaign (http://www.worldenergycity.com/cm_Print.asp?cmPageID=253) to market this relatively new concept to community members and business. While one of the major focuses of Halifax's smart growth strategy is smart economic growth, the concepts of smart development and community planning are also key. (Smart Growth Principles for Halifax Regional Municipality. http://www.fivebridgetrust.ca/docs/WRWEO_01.pdf. Accessed: November 10, 2003; Healthy Growth in HRM. http://www.fivebridgetrust.ca/docs/

TrustHRMsubmission.pdf. Accessed: November 10, 2003; Frank Palermo: Vision for Urban Growth in Halifax Regional Municipality. http://www.atlanticplanners.org/whatnew/conf99/palermo.HTM. Accessed: November 15, 2003)

Prince Edward Island

Green Home Visit Project

Throughout the summer of 1999, the Southeast Environmental Association and the Bedeque Bay Environmental Management Association performed green home visits to homes around Prince Edward Island. The main objective of the project was to reduce in-home energy and water consumption and increase biodiversity. The long-term, ultimate goal of the project was to mitigate climate change and maintain or improve local water and air quality.

Staff of the Southeast Environmental Association would visit homes and perform an assessment the homes and property. Information on various green home topics were provided to homeowners. Attention during the home visits was paid to:

- Where homeowners could conserve water,
- Where homeowners could conserve energy,
- Ways to improve indoor air quality,
- Alternative lawn and yard care methods, and
- Auto efficiency.

Where homeowners were willing to pay a fee, a more detailed assessment of the home was completed, and the homeowner was provided with coupons and a native tree or shrub to enable them to landscape their yard with native plants.

While this was a successful program, its funding was for the summer of 1999 (Southeast environmental association. http://www.seapei.ca/greenhome.aspx. Accessed: November 18, 2003).

United States

Washington and Oregon

Green Spaces Program, Portland and Vancouver, Washington

An effort to protect green space in the Portland/Vancouver region began in the 1980s. Non-profit organizations, local governments, and community members came together and formed a coalition for this effort. The Green Spaces Program focuses on environmental education, habitat restoration, public outreach, and regional planning.

The program has evolved since its inception, changing with funding and community interests. The goal of the program was initially to support natural area inventories and mapping that was then used to develop a strategic conservation plan for the region. In 1995, the community voted to acquire an extensive network of trails and green space. There are currently a variety of regulatory and non-regulatory tools being created to help protect green spaces, water quality, floodplains and fish and wildlife habitat.

As with the GVRD, the Portland/Vancouver region is located along the Pacific Flyway for migratory birds, making green space protection and restoration very important. There is a need to find balance in this region between urban sprawl and habitat conservation.

The Green Spaces Program is one that can be applied to protect fish and wildlife habitats in all urban areas. The activities that are part of this program help to engage local communities through restoration projects or educational activities (US Fish and Wildlife Service. Green Spaces Program. http://oregonfwo.fws.gov/greenspaces/gs-program.htm).

Vine Street Revitalization, Seattle

Vine Street is located in downtown Seattle's Belltown neighbourhood. It is a one-way street that sees a relatively low volume of traffic each day. This neighbourhood is highly developed and is characterized by a lack of green space. Residents of the neighbourhood saw a need, and opportunity, to extend an existing community garden and, in turn, convert the length of Vine Street into green space. The two guiding principles of the Vine Street Revitalization are to create a pedestrian friendly environment and to promote nature in the city.

The Vine Street Revitalization began in 1997 and encompasses eight city blocks. In addition to expanding the existing community gardens and planting native plants to encourage wildlife habitat, the Vine Street project has a stormwater management component. Stormwater will no longer run through storm sewers, but will be managed naturally through filtration bioswales. There are also plans to collect rainwater from local roofs and move this water through an aqueduct to a runnel. This runnel will then act as a wetland, and plants will be planted with the goal of absorbing toxins and cleaning water.

The overall cost of the Vine Street Revitalization is fairly high, and the ultimate plans for the area may take a few years realize. This idea behind this project is that it will continue to develop, and that the revitalization of Vine Street will last for future generations to see (Johnson, 2002).

California

Santa Monica Bay, Santa Monica

The Santa Monica Bay Watershed is located just outside Los Angeles and is made up of a variety of habitats that support fish and wildlife populations that, in turn, provide humans with many economic, recreational, scenic, and educational benefits. This watershed is made up of unique and interrelated habitats that have been adversely affected by urban development and population growth in the area over the years. Water quality, habitat integrity and biodiversity have all been affected in this watershed just outside of Los Angeles.

A variety of government agencies and environmental organizations are currently working on a number of projects in the Santa Monica Bay Watershed. These projects are solutions to habitat degradation, which have a positive impact on the great ecosystem.

In March 2000, Proposition 12 *(the Safe Neighbourhood Parks, Clean Water, Clean Air, and Coastal Protection Bond Act)* was passed by California's voters. This measure provided $25 million to implement actions identified in the Santa Monica Bay Restoration Plan. The following are examples of some of the restoration projects that have already taken place.

Madrona Marsh Restoration Project

The City of Torrance and the West Basin Municipal Water District plan to work together to rehabilitate Madrona Marsh. This project's goal is to expand open-water habitat, establish riparian vegetation, remove non-native species, cultivate native scrub plants, and distribute recycled water to the site. It will also serve as a model for other similar urban storm water management and wetland restoration and enhancement projects. The total funding for this project is $780,000.

Las Virgenes Creek Restoration Project

The city of Calabasas plans to restore Las Virgenes Creek by removing over 250 linear feet of concrete that lines a portion of the creek. This creek is an important tributary to Malibu Creek and Lagoon. Stream banks will be stabilized using a bio-engineered revetment and replanted with native vegetation.

Topanga Creek Lagoon and Lower Watershed Restoration Plan

The Resource Conservation District of the Santa Monica Mountains and the Topanga Creek Watershed Committee will develop a Lagoon and Lower Watershed Restoration Plan that will fully integrate the recreational, water quality improvement, and habitat enhancement needs of the area. The total funding for this project is $298,760.

Removal of Rindge Dam

As part of the ultimate goal of restoring Malibu Lagoon and Creek this project will assess the environmental impacts, alternatives and costs of removing Rindge Dam. Total funding for this project is $200,000.

Restoration of Natural Resources in Rocky Intertidal Habitats in Santa Monica Bay

This project will identify the causes of degradation in the Bay's intertidal habitats. Factors such as poor water quality, over-harvesting of intertidal species, and abuse by humans will be studied. Researchers will identify, evaluate, and recommend practical restoration techniques. The total funding for this project is $88,421.

Malibu Creek Habitat Enhancement: Removal of Arundo donax

The Mountains Restoration Trust plans to eradicate Arundo donax (Giant reed) in the Malibu Creek watershed. The project includes mapping, monitoring, and removal of the invasive species. The total funding for this project is $189,000.

Solstice Creek Restoration

The National Park Service plans to eradicate False caper (*Euphorbia terracina*) and other invasive perennial weeds, and restore native plants along five kilometres of Solstice Creek. The focus of the project is riparian understory restoration and enhancement, however non-native ornamental trees will also be removed. This project is linked to a multiagency program to restore Solstice Creek as viable southern steelhead trout habitat. The total funding for this project is $55,000.

Kelp Restoration Project

Santa Monica Bay Keepers plans to restore kelp to its historic acreage throughout the Santa Monica Bay by implementing a kelp restoration project, partially funded ($50,000 of $367,000) by the Bay Project. This project includes designing and constructing a kelp mariculture system in which kelp can be cultivated and transplanted to any of six selected restoration sites along the Malibu coast. The total funding for this project is $50,000.

Lower Topanga Canyon Arundo and Non-native Plant Eradication

The Mountains Recreation and Conservation Authority plans to eradicate Arundo donax and other non-native plants along a one-mile-long reach of Topanga Creek using a combination of techniques. The total funding for this project is $180,000 (www.santamonicabay.org/site/programs/layout/habitat.jsp).

Colorado

Tributary Greenways Program, Boulder

The Tributary Greenways Program is an extension of Boulder, Colorado's Greenways Program to riparian areas. The purpose of this program is to inventory, preserve and restore riparian areas for wildlife habitat.

The program is necessary for the health of Boulder's watercourses. After years of urban development, the municipality's riparian areas have been greatly altered. These habitats can contribute greatly to the overall ecosystem health of the community. While these systems may not be restored to their original state, or to support high levels of diversity, through restoration, they will be able to support a greater diversity of plants and animals. Specific and significant habitat features can also be protected or restored.

The first phase of the program, described below, is to assess the conditions of the terrestrial/riparian habitat of each creek and record the information. The main objective of this inventory is to collect baseline information on the condition of the city's riparian areas that will help guide restoration, protection and other management activities.

The methodology for the assessment of Boulder's riparian habitat was completed in 1999. The components of these habitats that will be assessed include: landscape factors such as area and width of the study area; biotic factors, such as plant species cover and richness and bird species richness; and human factors, such as adjacent land use and intensity of use, as well as ownership. In total, 141 sites were assessed along 13 creeks within Boulder. The field assessments took place between June and August 1999, and reports were then completed by ecologists contracted by the city. The following are themes that resulted from this study:

- Mature stands of Boulder's riparian forests are dominated primarily by native and introduced deciduous species, including crack willow, plains cottonwood, green ash, and Russian olive.
- Native species are not common components of the understorey, except in wetland areas.
- Introduced species and noxious weeds are present at almost all study sites.
- Common garden plants have become established at many of the study sites due to the proximity of residential areas.
  The riparian areas are often constrained by adjacent land uses.
- The conditions of the creeks studied can be improved through interdepartmental agreements, landowner education and stewardship programs.

Species richness and species weighted values were used to assess bird species diversity. The results varied between study sites, and richness was predictably higher in open spaces and undeveloped areas.

The results of these studies were then used to help guide habitat restoration and protection activities in Boulder's riparian areas (www.ci.boulder.co.us/publicworks/depts/utilities/projects/greenways/riparian.htm).

Denver

Xeriscaping

Xeriscaping is a concept of gardening that produces attractive, sustainable landscapes that conserve water. It was first introduced in the 1970s by the Denver Water Board (Pettinger 1996) and has since spread throughout Canada and the United Sates. This practice is based upon seven sound horticultural principles.

- Planning and design — it is important to consider things such as soil type, slope, sun exposure, and existing vegetation.
- Turf alternatives — most of the water is used to maintain grassy areas that are often under utilized. Turf areas should be evaluated and under-used ones replaced with plants that require less water.
- Zoning of plants — plants should be clumped by their needs (water, sun, soil, etc.), and plants with low water requirements should be used most often.
- Using soil amendments — these can improve root growth, water retention and water penetration.
- Using mulches — they reduce evaporation and weed growth, and help prevent soil temperature fluctuations.
- Efficient irrigation — not all plants need the same amount of water; therefore, plants should be watered according to need, not a schedule. Irrigation systems that keep water lower to the ground and have bigger drops also reduce the amount of water lost to evaporation.

- Appropriate maintenance — xeriscaping does not mean neglect. Proper pruning and weeding are important. However, landscapes adapted to the local environment will need less maintenance, less water, less fertilizer, and less chemical pesticides.

This movement has been especially popular in some of the drier parts of the United States. Municipalities in Colorado, Arizona, and Florida all have programs to encourage residents to use these practices. They offer free or inexpensive seminars, maintain websites with local information, have printed information available upon request, and maintain demonstration gardens. One of the biggest challenges is fighting the stereotypes associated with xeriscaping. It is not just a dry landscape made up of rocks and cacti. It can be lush and beautiful, using both native and non-native plants.

It is estimated that 50% of the water used in the home is for landscaped areas. Using the above principles can reduce water consumption by 30–80%.

Illinois

City Hall, Chicago

In April 2000, the US Environmental Protection Agency's Urban Heat Island Initiative created a 38,800 square foot semi-extensive green roof at Chicago's City Hall. This installation cost more than US$1 million. The main goal of this project is to highlight the benefits of green roofs in their capacity of moderating summer temperatures inside the ultra-urban environment.

Features of this green roof project include:

- a 3.5-inch deep "extensive" system,
- a 24-inch deep "intensive" landscape islands and
- a drip irrigation system, collecting water, in part, from adjacent buildings.

The green roof project at Chicago's City Hall was completed in the fall of 2001. It will be monitored for plant survival, stormwater management and other environmental benefits (Low Impact Development Centre, Inc. http://www.lid-stormwater.net/greenroofs/greenroofs_commercial.htm).

Chicago Wilderness: A Regional Nature Reserve

Chicago Wilderness refers to a place — natural communities in northeastern Illinois, southeastern Wisconsin and northwestern Indiana, and a collaboration — and to the organizations that protect and restore the remnants of natural land that occur here, including the best concentrations of prairies and oak woodlands in the world. About 200,000 acres (77,000 ha) of Chicago Wilderness are formally protected, but most of this is disturbed by grazing, fire suppression, invasives, drainage and other human activities. Nevertheless, they are important habitats for rare plants and animals.

The organizations include: forest preserve districts, park districts, state departments, zoos, gardens, nature conservancies, universities, municipalities, Audubon societies, foundations and private business. Together they have produced a Biodiversity Recovery Plan that includes visions for the entire landscape, ecosystems, individual species and the people of the region. The key recommendations of the plan are (Chicago Region Biodiversity Council 1999):

1. Land that is already preserved for conservation must be managed in such a way that biodiversity is protected and restored.
2. More land with existing or potential biodiversity benefits must be preserved.
3. Water resource management programs, policies and regulations that are sensitive to sustaining native natural communities need to be developed and implemented.
4. More monitoring and research needs to occur to better understand the requirements for biodiversity conservation and ecological restoration
5. Increase citizens' awareness of the importance of biological conservation and the opportunities to participate in programs and efforts that enhance biodiversity.
6. Local and regional development policies should reflect the need to restore and maintain natural areas and biodiversity.

The recovery plan also includes a detailed list of actions that can be taken by county and local governments and agencies, state agencies, federal agencies, businesses, owners of large tracts of open land, owners of farmland and individuals. The Chicago Region Biodiversity Council has produced a number of documents to help support these actions, including Protecting Nature in Your Community (2000), a resource guide with successful examples of community initiatives.

A study of public attitudes toward ecological restoration in metropolitan Chicago suggested that public attitudes towards ecological restoration may be greatly influenced by familiarity with restoration projects and the proximity of such projects to where they live (Bright et al. 2002). Chicago Wilderness is an effective positive influence on cognitive (perceived outcomes, value orientation, objective knowledge), affective (emotional) and behavioural factors that determine a community's attitude towards ecological restoration.

Indian Bounday Prairies — Chicago

The Indian Boundary Prairies are a cluster of four prairies located south of Chicago near the junction of US Route 57 and Interstate 294. At more than 300 acres, they are the largest remaining example of high quality grasslands in Illinois. Due to their size and natural state, these prairies contribute greatly to the region's biodiversity, and serve as a store of genetic diversity for the future. The prairies are protected through ownership by the Nature Conservancy, Northeastern Illinois University and the Natural Land Institute, and the Friends of the Indian Boundary Prairies help with caring for the prairies and educating the public.

Efforts to protect the Prairies began in the 1960's when biologists from the Chicago area began to survey the Prairies and recognized their ecological importance. In 1971, 60 acres were donated to the Nature Conservancy, and the protection of this ecosystem began. One hundred acres of this ecosystem remain unprotected and are threatened by development.

A Northeastern Illinois University professor has successfully introduced Franklin's ground squirrel back into this ecosystem. These squirrels were original inhabitants of the Indian Boundary Prairies and became extinct through development and the introduction of non-native plants to the region.

The prairies are open to the public for hiking, bird watching and other activities (www.nature.org/wherewework/northamerica/state/illinois/preserves/art11119.html.

Pennsylvania

Fencing Academy, Philadelphia

The Fencing Academy in Philadelphia planted a 3,000 square foot meadow-like rooftop. This green roof initiative was created during the spring of 1998. This garden was part of a retrofit for the building, with the main goal of restoring the pre-development hydrology of the site. These areas were planted with a variety of *Sedum* varieties that have now formed a meadow-like setting.

The roof cover is only ¾" thick and weighs less than 5 pounds/square foot (lb/sq) up to a maximum of 17 lb/sq when fully saturated. This green roof has been continuously monitored for temperature since the project was completed. These temperature readings are then compared with adjacent unvegetated roofs. The daily temperature fluctuations of a bare roof are 90°F (50°C) compared with 18°F (10°C) to a green roof (Low Impact Development Centre, Inc. http://www.lid-stormwater.net/greenroofs/greenroofs_commercial.htm).

Florida

South Florida Ecosystem Restoration and Sustainability Project

The South Florida Ecosystem Restoration and Sustainability (SFERSP) project consists of nearly 200 environmental restoration, growth management, agricultural, and urban revitalization projects, programs and initiatives. These activities are aimed at making south Florida a more sustainable region. The main goals of this project are to restore the natural hydrology of south Florida, enhance and recover native habitats and species, and revitalize urban core areas to reduce the outward migration of residents to the suburbs and improve the quality of life in these core areas. This in an ongoing project, with plans to continue over the next 50 years.

SFERSP extends from the Chain of Lakes south of Orlando to the reefs surrounding Fort Jefferson, which is southwest of Key West in the Florida Keys. The project area includes both terrestrial and aquatic habitats, as well as the large urban areas of Miami, Fort Lauderdale, West Palm Beach, Fort Myers and other growing communities in south Florida. This greater ecosystem covers approximately 10,800 square miles (28,000 square kilometres) and includes 11 major physiographic provinces (similar to British Columbia's ecoregion classification system).

The restoration effort is a partnership between many organizations, including federal departments and agencies, state commissions and agencies, American Indian tribes, counties, municipalities, industry, and commercial and private sectors as well as special interest groups. Restoration activities are guided and coordinated by a task force involving members from many of the above organisations.

The restoration projects are guided by three main strategies: adaptive management, innovative management and action. The task force members and other participants recognize the need to continually collect data and refine project plans. Feedback and alternative project plans are important components of the SFERSP, and ensure accountability and flexibility. These restoration projects are based on scientific research and also incorporate innovative, holistic and outcome-oriented management plans. Ambition and action are also key components of these projects. These strategies guide the SFERSP and ensure its success.

These ultimate success of the restoration projects that are part of the SFERSP will take years or decades to realize. From the progress that has been made to date, plans will be continually updated and methods improved. One final goal of the SFERSP is that the methods and strategies used in south Florida can be applied in other regions (www.sfrestore.org/documents/success/02.htm).

Europe

European Cities and Towns towards Sustainability Campaign

Eighty percent of citizens of the European Union live in urban areas. This fact is one of the reasons urban sustainability is a priority with the political authorities of the EU.

The European Sustainable Cities and Towns Campaign was started in 1994 with the adoption of the Charter of European Cities and Towns towards sustainability or the Aalborg Charter. This charter encourages its participants to work towards sustainability in urban settings and to create a Local Agenda 21 plan. Originally 80 local authorities signed this charter; however, now there are more than 1650 members.

In 1996 the Lisbon Action Plan translated the ideals of the Aalborg Charter into concrete steps that could be taken by the local authorities. The Campaign acts to provide support and disseminate information to the municipalities by promoting sustainable development at a local level as recommended by the Earth Summit of 1992.

Below are some concrete examples of what European cities are doing towards sustainability (europa.eu.int, www.sustainable-cities.org, www.eaue.de/winuwd/list.htm surbandatabase).

Slagelse, Denmark is a small city of 35,000 people located 80 km from Copenhagen. It decided to undertake an urban renewal of its dilapidated downtown core that consisted of 8 blocks, 770 buildings, and 1,500 inhabitants. The goals were to improve the environment and to improve quality of life for its citizens. This was achieved by:

- Use of ecological building materials.
- Addition of solar panels and use of solar energy.

- Separation of waste products: paper, glass, and compostable wastes. Composting of household wastes achieved almost 100% participation.
- Creation of open space for all activities. They created "green rooms" with different themes, such as play, rest, communication, sun, shade, walking, gardening, and animal keeping and divided these rooms with ponds and espaliers.
- Closed one street to traffic and created a pedestrian zone.
- Provide artistically designed electrical outlet posts for electrical cars.

This area is now a vital part of the city. The project succeeded because of political support and public participation.

STATTAUTO is a car-sharing company started in Berlin, Germany. Car sharing is where a group of people own a fleet of vehicles that can be signed out for use when required.

Research has shown that people use their cars less than one hour per day in Germany. It is expensive to have a car and car-sharing is one way to cut down on costs (both financial and environmental) while not losing all of the independence that comes with car ownership. This particular car-share started out as a research project and its popularity caused it to take off. In 1996, there were over 4,000 members. They contributed to an annual reduction of 510,000 car kilometres and 80.32 tonnes of CO_2 emissions, while at the same time creating new employment opportunities.

In 1993 Bologna, Italy participated in the Urban CO_2 Reduction Project of the International Council for Local Environmental Initiatives, whose goal is to reduce carbon dioxide emissions by 29% by 2005. The project focused its measures in the areas of increased use of natural gas and renewable energies, the identification of energy saving potentials, the planning of a comprehensive traffic and transport policy, and increased planting of urban vegetation.

Some examples of ways to achieve this goal are:

- Municipal-owned buildings would be upgraded using energy efficient and environmentally friendly systems.
- Hospitals and universities would be test cases for energy efficiency.
- Information campaigns would be run to encourage the public to become more energy efficient.
- Public transport should have the priority in traffic policy and measures to reduce private car use should be implemented.
- A new waste incineration plant should be built and alternate energies from wastes should be pursued.
- An increase of 10% in biomass by additional plantings in public spaces (50, 000 new trees and 100,000 new shrubs).

Spain

Madrid

In response to the Earth Summit 1992, the increasing numbers of people living in cities, and increasing levels of environmental consciousness the municipal government took action to make an effort to preserve city life. They believed that urban and marginally urban environments must be understood and managed as ecological systems, indigenous species of plants and animals should be reintegrated into the urban environment, and planning policies should provide for and preserve links to vegetated areas both within and out of the city. They embarked on a study to naturalize the large surfaces of urban buildings. Some of the goals of this project are:

- Reduction of dust load and emissions
- Reduction of traffic noise
- Optimizing the atmospheric humidity by accumulation of rainwater in the plant layer on roofs
- Reduction in rain water to sewer systems
- Improvement of the energy efficiency of buildings by saving on heating and cooling costs
- Reduce roof damage by reducing ultraviolet radiation on the surface of the roof
- Preservation of groundwater reserves
- Increase biodiversity

As the previous examples illustrate, activities that contribute to nature's services can take many forms. These are just a few of the thousands of projects that are have taken place and are ongoing. As shown above, restoration projects can be: large or small scale; spearheaded by a municipality, regional districts, stewardship groups, concerned citizens, or a coalition of several organizations; are implemented by paid or volunteer workers; and can have varied degrees of expenses. While these projects are very different in scope, they also have many things in common, the most important being their contribution to ecosystem function and diversity.

18

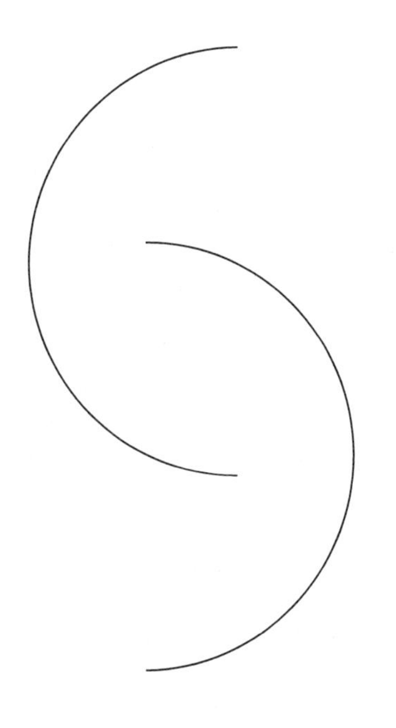

A New Urban Reality

The loss of important habitat due to urbanization is considerable. In the United States from 1959 to 1982, 22 million acres of land were converted to urban and other developed land uses, an increase of 45% (Heimlich and Anderson 1987). In the Lower Mainland of British Columbia, land that has been converted to agriculture (or other similarly cultivated landscapes) has only 50% of the average net primary productivity of original forested ecosystems and urban landscapes have only 13% (Healey 1997). The natural ecosystems surviving are often fragmented and isolated with reduced biodiversity and a loss in viability. Fragmentation also facilitates the spread of invasive alien species (Wilcox and Murphy 1985). Because cities are often situated in highly productive the ecosystem fragments are far more important than their limited size and disturbed plant life might initially suggest (Schaefer 1994).

We sleep, eat, work and play in an ecosystem, which can be defined as life, earth, air, and water interacting with one another in certain special ways. We do not normally meet most of the other participants in this system, and we are unaware of the impacts we have on them, or that they, in turn, have on us. But this lack of awareness is not surprising. We live in the city in a cultural context that is totally foreign to biological evolution. We surround ourselves with objects, machines, and dwellings that we have manufactured and that bear no resemblance to the raw materials from which they were made. We often spend long periods of time in parts of the city where nature is nonexistent or necrotic, diseased and in poor health.

Many of us have only recently become aware of our dependence on nature. However, this awareness often seems to translate into a concern with global rather than local issues. We talk about global impacts, global problems, a global community: we identify more with the globe, the biosphere, and its associated problems such as acid rain, the greenhouse effect and tropical rain forest destruction, than we do with our immediate surroundings.

This global focus, however, often does not convey a concern for the local ecosystem — is suggests that we only should act locally because of a global impact. The phrase "Think globally, act locally" sums up this point of view. Things are only worth doing if they can change the world.

From a different perspective, it has been said that we do not need a global village, but rather a globe of villages. By acting locally, we become aware of, and as a result, become accountable for our own actions. We protect our local ecosystem for its own sake. This eventually has a global impact, but the orientation is significantly different and may be more successful in effecting community change and, ultimately, the protecting of the biosphere.

By acting locally with a local impact focus, we are dealing with matters where we can have an immediate impact, today, alone, unaided. Every plant matters. Also, we are dealing with issues where the average person can effect political change (and through this, changes in the environment), locally, face-to-face, with the elected representatives of the community. With only a little effort, the individual can act in an empowered group. Global changes in politics and world views are necessary, and will continue to be pursued. The means for these changes are, however, very costly in time and money, and inaccessible to most people. Many people become discouraged, give up, and adopt a passive lifestyle.

We destroy parts of our natural ecosystem to house and feed ourselves, and to drive and park our cars. Little by little, we erode and destroy the system that supports us. Pollution problems indicate that our numbers and activities already exceed the capacity for the ecosystem to support us.

Restoring urban habitat often means focusing on the greater collective impacts of numerous small-scale projects.

High concentrations of people with the current North American lifestyles cannot be supported by the natural ecosystems in which they occur. Slowly, but certainly, we are poisoning the local ecosystems that we live in. We threaten the future of our cities, and all life around us — the very ingredients that maintain us practically and spiritually. Our environmental problems and issues are not considered life threatening yet, but they certainly do affect the quality of our lives.

Managing for biodiversity in urban situations is different from managing for biodiversity in wilderness areas. A new paradigm is emerging with a focus on target species, indicator species, patch sizes, adjacency of nodes and connectivity. When managing for urban biodiversity, we manage both for wildlife and for nature's services, particularly for how these services can benefit air and water quality and provide food or otherwise benefit people.

What are the differences in managing for wildlife versus managing for nature's services? Managing for wildlife may involve considerations of patch characteristics, effect distances, all waterways, the properties of understorey vegetation. Managing for nature's services, in addition to including areas managed for wildlife, may include considerations of the aggregate impact provided by all lawn areas, street trees, pocket parks, spaces without understorey, rooftops and other marginal habitats.

How do we make urban habitats with greater ecological integrity? Many biologists are concerned that enhancing urban habitats will simply create habitat "sinks", luring animals from the habitat "source" in the surrounding wilderness to face almost certain death or, at least, much reduced reproductive success in urban areas. How can we improve urban biodiversity without creating such habitat sinks?

The city (i.e., people, their lifestyles and technology) needs to get along with nature, and nature in turn, must come into the city. This is the new city plan for the future — the Green City. Recycling, bicycling, composting, walking, enjoying, gardening with native vegetation, and much more, all play their part. New urbanism, a design movement that gained momentum in the late 20th century, promotes walkable neighbourhoods that support the Green City Approach and regional planning for green space.

What does this all mean to our ecosystem in terms of energy flow, the recycling of nutrients, and the generation of wastes? What do we do to our ecosystem, and what does it do to us in response? What are the parts of the system? What are the particular sets of relationships? What changes are happening in our current lifestyles, and how will we make decisions about the future? The answers to these questions will be more apparent when we are more successful at distinguishing "quality of life" from "standard of living". We will very likely be happier and healthier if we enable nature to play a larger role in our lives, here in the cities, where most of us live.

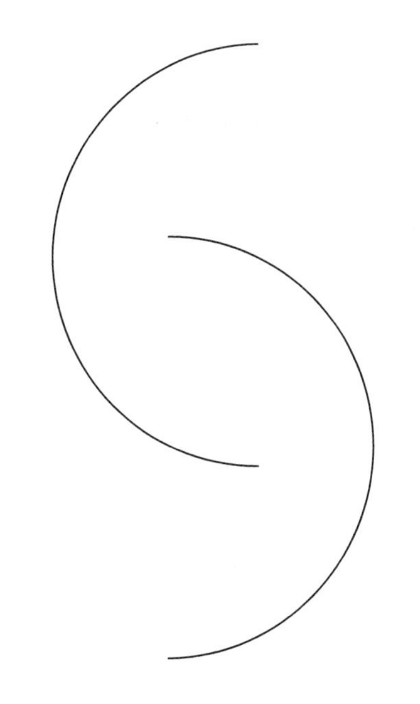

Appendix

Some organizations with a major aim to study nature in the city and increase urban biodiversity.

Canada

Douglas College Institute of Urban Ecology
P.O. Box 2503
New Westmintser, BC V3L 5B2
www.douglas.bc.ca/iue
Tel: 604-527-5224
Fax: 604-527-5095
email: iue@douglas.bc.ca

Environmental Research and Studies Centre
University of Alberta
3-23 Business Building
Edmonton, Alta, T6G 2R6
www.ualberta.ca/~ersc
Tel: 1-780-492-5825
Fax: 1-780-492-3325
e-mail:beverly.levis@ualberta.ca

Urban Ecology Center
C.P. 473
Succ. Place du Parc
Montreal, Que H2W 2N9
www.urbanecology.net/UEC/indexuec.html
Tel: 514-282-8378
Fax: 514-985-9725
sodecm@web.net

United States

Urban Ecology
414 13th St. Suite 500
Oakland, CA 94612
www.urbanecology.org
Tel: 510-251-6330
Fax: 510-251-2117

Center for Urban Forest Research
c/o Dept. of Environmental Horticulture
University California at Davis

Suite 1103 One Shields Avenue
Davis, CA 95616
wcufre.ucdavis.edu
Tel: 530-752-7636
Fax: 530-752-6634
email: cufr@ucdavis.edu

Urban Ecology Center
Riverside Park
2808 N. Bartlett Avenue
Milwaukee, WI 53211
http://my.execpc.com/~uec/
Tel: 414-964-8505
email: uec@execpc.com

Center for Urban Ecology
4598 MacArthur Boulevard NW
Washington, DC 20007-4227
www.nps.gov/cue/cueintro.html
Tel:202-342-1443
Fax: 202-282-1031

Urban Ecology Institute
Boston College
225 Higgins Hall
Chestnut Hill, MA 02467
www.bc.edu/bc_org/research/urbaneco/
Tel: 617-552-0592
Fax: 617-552-1198
email: heather.langford@bc.edu

Urban Ecology Research Laboratory
Box 355740, Gould 410
University of Washington
Seattle, WA 98195
www.urbaneco.washington.edu
Tel: 206-616-9379
email: malberti@uwashington.edu

Buffalo Institute of Urban Ecology Inc
1645 Statler Towers
107 Delaware Avenue
Buffalo, NY 14222
www.nfwhc.org/groups/buffalo.htm
Tel: 716-845-6993
email: jburney@apollo3.com

Hixon Center for Urban Ecology
Yale University
285 Prospect Street
New Haven, CT 06511
www.yale.edu/hixon/
Tel: 203-432-5104
email: hixoncenter@yale.edu
Urban Ecology Coalition
P.O. Box 7423
Minneapolis, MN 55407
www.sustainable.doe.gov/measuring/meneighbor.shtml
Tel: 612-869-8664
email: kmeter@igc.apc.org

North St. Paul Urban Ecology Center
University of Minnesota
Department of Landscape Architecture
89 Church St. Southeast
Minneapolis, MN 55455
www.snre.umich.edu/nassauer/NSP/NSPCover3.html
Tel: 612-763-9893
email: nassauer@umich.edu

Center for Urban Restoration Ecology
Rutgers University and Brooklyn Botanical Garden
1 College Farm Road
New Brunswick, NJ 08901-1582
www.i-cure.org/
Tel: 732-932-8165
Fax: 732-932-2972
email: cure@aesop.rutgers.edu

Center for Urban Ecology and Sustainability
University of Minnesota
219 Hodson Hall
1980 Folwell Avenue
St. Paul, MN 55108
www.entomology.umn.edu/cues/cuesindex.htm
Tel: 612-625-7044
email: krisc001@tc.umn.edu

Brooklyn Center for the Urban Environment
The Tennis House, Prospect Park
Brooklyn, NY 11215-9992
www.bcue.org/index.htm
Tel: 718-788-8500
Fax: 718-499-3750
email: info@bcue.org

Institute of Ecology
University of Gerogia
Ecology Building
Athens, GA 30602-2202
www.ecology.uga.edu/
Tel: 706-542-2968
Fax: 706-542-4819
email: rosemond@sparc.ecology.uga.edu

New Ecology Inc.
Greenworks Buidling
160 Second Street
Cambridge, MA 02142
www.newecology.org
Tel: 617-354-4099
Fax: 617-354-4098
email: info@newecology.org

Tri-City Ecology Center
P.O. Box 674
Fremont, CA 94537
www.tricityecology.org/index.html
e-mail: tcec@tricityecology.org

Committee on Urban Environment
Minneapolis Planning Department
350 South Fifth Street B Rm. 210
Minneapolis, MN 55415-1385
www.ci.minneapolis.mn.us/citywork/planning/sections/cue/contact.asp
Tel: 612-673-3014
cue@ci.minneapolis.mn.us

Central America

Instituto Nacional de Biodiversidad
P.O. Box Apdo. 22-3100
Santo Domingo de Heredia, Costa Rica
www.inbio.ac.cr/en/
Tel: 506-244-0690
Fax: 506-244-2816
email: askinbio@inbio.ac.cr

Europe

Centre for Urban and Regional Ecology
School of Planning and Landscape
University of Manchester
Manchester, UK M13 9PL
www.art.man.ac.uk/PLANNING/cure/
Tel: 44-(0)161-275-6920/38
email: cure@man.ac.uk

Institute of Agricultural and Urban Ecology Projects
Humboldt University Berlin
Invalidenstr. 42
D-10115 Berlin, Germany
www.hu-berlin.de/forschung/aninst/aninste.htm
Tel: 49-30-2093-8683
email: r.sauerbrey@agrar.hu-berlin.de

Danish Centre for Urban Ecology
Jaegergardsgade 97
P.O. Box 5095
DK-9100 Arhus C, Denmark
www.dcue.dk/uk/intro-uk.htm or www.dcue.dk/Default.asp?ID=7
Tel: 89-40-58-80
Fax: 89-40-58-84
email: dcue@dcue.dk

ECOPLAN: Ecology and Urban Planning
Department of Ecology and Systematics
P.O. Box 17
FIIN-00014 University of Helsinki
Helsinki, Finland
www.helsinki.fi/~niemela/ecoplan.html
Tel: 09-1917391
Fax: 09-1917301
email: jari.niemela@helsinki.fi

International Institute for the Urban Environment
Nickersteeg 5
2611 EK Delft
The Netherlands
http://www.urban.nl/
Tel: 31-15-262-3279
Fax: 31-15-262-4873
email: iiue@urban.nl

European Academy of the Urban Environment
Bismarckallee 46-48

D-14193 Berlin, Germany
http://www.eaue.de/default.htm
Tel: 49-30-8959-990
Fax: 49-30-8959-9919
email: am@eaue.de

Australia

Centre for Urban Ecology
105 Sturt Street
Adelaide
Tandanya Bioregion SA,
5000 Australia
www.urbanecology.org.au
Tel: 61-8-8212-6760
Fax: same
email: urbanec@urbanecology.org.au

Australian Research Centre for Urban Ecology
c/o School of Botany
University of Melbourne
Victoria 3010, Australia
arcue.rbg.vic.gov.au
Tel: 61-03-8344-0146
Fax: 61-03-9347-5460
email: arcue@rbg.vic.gov.au

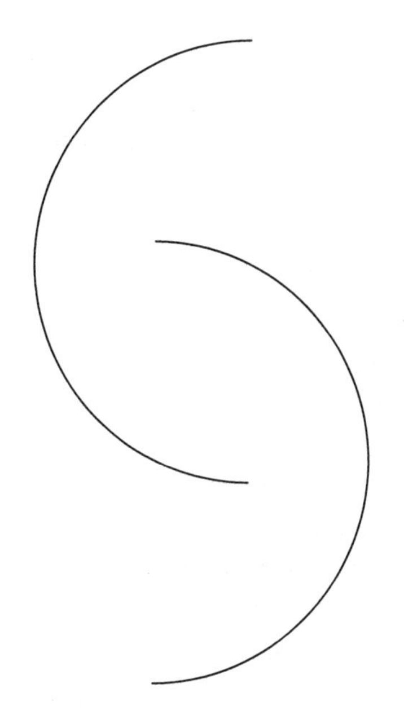

Glossary

abitoic nonliving components of an ecosystem, such as soil and water

acclimatize when an organism alters physiologically in response to being exposed to a different environment

acidic having a pH of less than 7.0 (which is neutral)

adaptive management incorporating changes into management as a result of feedback on the progress of a management strategy

aerobic in the presence of oxygen

Agenda 21 arising from the Earth Summit, a plan for action, objectives, activities and means of implementation of recommendations to address environmental issues

aggregate sand and gravel used as structural material, or to make concrete and asphalt

albedo solar radiation reflected from the Earth's surface

alkaline having a pH of more than 7.0 (same as basic)

alpine trough an elongated plant container with species normally found at high elevations

ambient pertaining to the immediate surroundings

anaerobic in the absence of oxygen

annual lasting for just one year

aquatic pertaining to water, either freshwater or marine

arborist a person trained in the cultivation, care and maintenance of trees and shrubs

asexual reproduction a new individual is formed from a part of, or an outgrowth from, an individual where there is no union of male and female gametes

Audubon Cooperative Sanctuary Program a certification process to recognize environmentally friendly designs and practices for golf courses

autotrophic a mixture of decomposed organic material and bacteria

bankful channel width the width of the channel up the bank that is devoid of permanent vegetation because it is scoured by heavier water flows at certain times of the year

bankful depth the depth of water in a stream or river during its highest water flow

basic having a pH of more than 7.0 (same as alkaline)

bat box a box of wooden slots used to provide bats with a roosting or hibernating site

bee block a solid wooden block with holes used to provide mason bees with a breeding and wintering site

benthic the region on, or in the bottom of, a water body, such as a stream, lake or intertidal area

best management practice (BMP) procedure for organizations, businesses and industries that promotes sustainable operations and minimizes environmental impact

biodiversity the variety of life, including the levels of individual, population, species and ecosystem

bioengineering the use of living-plant material to stabilize a slope or perform some other function traditionally done using materials such as rocks, cement, iron bars or concrete blocks

biological inventory a scientific assessment of the species, distribution and abundance living in an area

biomass the weight of organisms in a given area

biome major community of plants and animals covering a large geographic area, for example taiga or tundra

bioregion an area of similar plant and animal life, topography and human culture

bioregionalism the practice of "living in place" supporting the local environment, economy and culture

biosphere the part of the Earth's surface and atmosphere that contains living organisms

biota organisms

biotic the living components of an ecosystem, such as plants and animals

blacktop areas paved with asphalt

bog mineral- and nutrient-deficient wetland with an acidic organic environment dominated by peat sphagnum moss

boulder a rock greater than 25 cm in diameter

bramble a thorny plant of the genus *Rubus* often forming dense thickets of tangled vines

branching network an open-ended pattern of pathways

brownfield land, such as a manufacturing plant or an industrial or commercial facility, with residual substances in the soil left over from the previous use of the site

bryophyte moss or liverwort

buffer zone an area around a habitat that mitigates disturbance to the habitat from its surroundings

bylaw a legal ordinance of a municipality or a standing rule that regulates the internal affairs of an organization or community

canine zone an area around the lower part of a tree trunk whose appearance is changed because of frequent urination by dogs

canopy the upper region of trees in a forest in the high branches

carbon sequestration the removal of carbon dioxide from the atmosphere by plants

carbon sink a reservoir for organic material that is unavailable for decay and, therefore, does not contribute carbon dioxide to the atmosphere

carrying capacity the maximum number of individuals from a species an environment can support over an extended period of time

charette a session of the general public and experts such as landscape architects to explore ideas for future community structure

circuit network a closed pattern of pathways

citizenship becoming informed about an issue and acting on that information

CITYgreen a program developed by American Forests to calculate nature's services provided by urban greenspace, including such benefits as: reducing stormwater runoff, air pollution reduction, energy conservation and increased wildlife habitat

climax community one can maintain itself indefinitely if left undisturbed

clone a population of genetically identical individuals created by asexual reproduction

cobble streambed material 5–25 cm in diameter

commensalism a symbiotic relationship where one species benefits and the other is unaffected

community all organisms of all species living in an area

competition a symbiotic relationship between two organisms or species where each would be better off without the presence of the other

comprehensive development zone a municipality creates customized zoning regulations with permitted uses, densities and desired amenities to deal with a complex site.

connectivity the ability or inability of wildlife to move between patches of habitat

Convention on Biological Diversity arising from the Earth Summit, established three main goals: the conservation of biological diversity, the sustainable use of its components, and the fair and equitable sharing of the benefits from the use of genetic resources

Convention on Climate Change a framework to reduce greenhouse gas emissions agreed upon at the Earth Summit in 1992

Covenant a form of land tenure where conservation covenant can be applied to a property that restricts public access in an environmentally sensitive area, or redevelopment of a private property

Crime Prevention Through Environmental Design (CPTED) an approach to community design where the architecture and landscape are used to reduce crime

culvert a pipe used to conduct water underground as, for example, under a road

daylighting exposing a stream that had been covered by development to the open by removing a covering of ground and restoring its structure

deciduous forest a forest of trees that lose their leaves seasonally to reduce water loss

dedication a form of land tenure where the management and maintenance of land can be dedicated to a local government

denitrification the conversion of nitrogen in organic molecules into atmospheric nitrogen

density number of organisms in a population expressed per unit area

density bonusing a developer may apply for extra units if they in exchange for protecting an environmentally sensitive area

detention pond a pond that temporarily keeps water on a site

detritus obtaining energy from other organisms by ingestion or absorption

directed attention fatigue tiredness due to increased concentration necessary to perform a task as a result of distractions or lack of interest

dissolved oxygen the amount of oxygen dissolved in water available to support aquatic life

Earth Summit an international meeting held by the United Nations in Rio de Janeiro in 1992 to address global environmental issues

ecocity a city in harmony with the bioregion, constructed on the principles of life, beauty, and equity

ecological footprint the amount of land it takes to support our needs and lifestyles

ecological rucksack the hidden inventory of materials, resources and impacts involved in producing a product

ecosystem the living and nonliving components of an area that interact according to a particular set of relationships

edge effect the increase in the number of species found in the area between two habitat types, or a decrease in numbers of a particular species inside a habitat patch due to outside disturbance

effect distance the distance from the source of a disturbance (such as traffic noise) to where it can disturb wildlife

embeddedness the degree to which stones in a streambed are buried by sediment

endangered species one that is threatened by extinction because of overharvesting or habitat destruction, especially those with low population sizes and low reproductive rates

environmentally sensitive area a biologically important area identified for natural importance and vulnerability to impact from human disturbance

epiphytic growing on branches and trunks of trees without deriving energy from them

estuary a semi-enclosed coastal body of water that has a free connection with the open sea, and within which sea water is diluted with fresh water derived from land drainage

ethnobotany the study of the cultural uses of native plants

eutrophic lake generally older, shallow, warm lakes with a low dissolved oxygen content and large amount of dead organic material

evapotranspiration the water lost through evaporation and by transpiration through plant leaves (*see also* transpiration)

Fatal Light Awareness Program (FLAP) a program to prevent birds from striking tall buildings encourages tenants to turn their lights out at night

feral having become wild after being in captivity

First Nations the original human population indigenous to an area

floodplain the low-lying lands adjacent to a stream or river that are naturally subjected to flooding

flowering plants plants of the class *Angiospermae* that are either in the subclass *Monocotyledonae,* such as grasses and tulips, or the subclass *Dicotyledonae,* such as carnations and all of the deciduous woody trees and shrubs

fragmentation the reduction of large patches of habitat into smaller ones

functional niche the sum of all of the interactions of a species with others in an ecosystem

garden city a city beautified with flowers and landscaping with a sensitivity to scenic views

generalist having broad tolerances for habitat, food, shelter and breeding sites

geographic information system (GIS) a computer-based system that uses data collected electronically to combine layers of information about a place

grassland an expanse of grass, possibly with some scattered trees

gravel small rocks 0.2–5 cm in diameter

grazer feeding mainly on leafy vegetation

green city a city that incorporates a strong connection with natural areas and ecological processes

greenfield agricultural lands planted with grasses that can provide valuable wildlife habitat, especially for hawks and owls

greenway a vegetated corridor for people or wildlife; it often includes a path for pedestrians and cyclists

gross primary productivity the total amount of sunlight energy fixed by plants

ground cover low herbaceous plants and crawling shrub growth covering the ground

groundwater water found beneath the surface of the ground that largely originated from water percolating from the surface or moving laterally through the soil

growth form a category of plants based on their morphological appearance, such as tree or shrub

habitat the land or water an organism occupies

habitat sink habitat features that attract a species, but in which there are insufficient resources for it to live and breed (resulting in death to the adults and/or young)

habitat source where a species can live, breed and thrive

hedgerow a closely-planted row of trees, shrubs, and perennials that are in a natural linear arrangement

heterotrophic obtaining energy from the sun through photosynthesis or from chemosynthesis

hole-nesting a species that normally builds its nest in a tree cavity

home range the area in which an animal moves over the course of a year

hydrology the distribution and circulation properties of water on a site

impervious surface a surface impermeable to water, such as a sidewalk or road

indicator species a species whose presence suggests a certain habitat, environmental condition, association of other species or other factor

infrastructure basic underlying features of a city, such as sewers and roads

integrated pest management (IPM) managing pests by using natural predators and cultivation practices as part of a program that may also include pesticides

interior species one that is intolerant of outside disturbance and requires a buffer from the edge of the habitat patch

interspecific between individuals of different species

intraspecific between individuals of the same species

invasive species non-native species that dominate an area they are introduced into

keystone species a species that has a dominant effect on a community and, if removed, will dramatically alter community structure

Kyoto Accord a protocol agreed upon in 1997 that committed industrialized countries to achieve quantified targets for greenhouse gas emissions

Landfill a site for disposing of solid waste where garbage is covered with mineral soil

large woody debris logs and stumps

leachate water leaving the landfill that seeped through the garbage, picking up toxic chemicals or the products of decomposition from the waste

leaf area index the area of tree foliage relative to the area of ground covered by the tree canopy

LEED Leadership in Energy and Environmental Design, a system of evaluating green-building design developed by the U.S. Green Building Council

limited disturbance clause specifying limited activities within a setback

limnetic zone open water beyond the littoral zone (*see also* littoral zone)

liquid waste human waste from toilets, sinks and drains

littoral zone region of shallow water with rooted plants

macrofauna larger animals, such as fish, amphibians, reptiles, birds or mammals

marsh wetland with mainly grassy vegetation, such as grasses, cattails, rushes, reeds and sedges

metapopulation the larger population composed of discrete smaller populations

microhabitat a small habitat on the scale of a log, stump or rock

micro-organism microscopic organism, usually bacteria or protists (single-celled organisms more advanced than bacteria)

mineral soil soi low in organic content

morphology the overall form of an organism

multidimensional niche, also hypervolume niche the sum of all conceivable connection between the species and the living and nonliving world

mutualism a type of symbiotic relationship where between two species where both benefit; can be obligative (e.g., lichen), where each species cannot exist without the other; or facultative (e.g., ants and aphids), where the relationship is not required for survival

mycorrhizae mutualistic fungal association with the roots of higher plants that improve the uptake of nutrients by the plant roots

native vegetation the vegetation in an area prior to human disturbance

nature's services the conditions and processes provided by nature that enable human survival

naturescaping landscaping in the ecological style, to be in harmony with nature

necrosis death or disease resulting from the death of part of an organism (tissue, organ) or ecosystem (species)

net primary productivity energy fixed by plants that is available to pass on to the next trophic level (consumers)

new urbanism an urban design movement that promotes walkable neigbourhoods with a diversity of jobs and housing that supports regional planning for green space

niche the role species play in an ecosystem

nitrogen fixation the process by which atmospheric nitrogen is converted into nitrates in the soil that can be used by plants

nitrogen fixing bacteria bacteria that can convert atmospheric nitrogen into nitrates and are found in nodules associated with the roots of certain plants

nutrient loading the buildup of nutrients from human activities in natural areas

official community plan (OCP) a description of how the community will approach land use, development and environmental issues in terms of its attitude and commitments

old field habitat abandoned agricultural fields overgrown with tall grasses and shrubs

oligotrophic lake lakes generally younger, tend to be deeper and stratified, with the lower temperatures usually resulting in a higher dissolved-oxygen content

organic soil soil rich in organic matter

ornamental plant a species chosen to adorn or embellish a site

Pacific Flyway following the western coast of North America, one of several major flight paths used by migratory birds during north-south migrations

parasite an organism feeding on another living organism (its host) without killing it

peat a mixture of undecomposed and partially-decomposed organic matter, especially Sphagnum moss, formed in bogs

perennial lasting for a number of years

permafrost a permanently frozen layer of ground found in the tundra

pesticide a chemical used to kill a pest than can be a herbicide (kills plants), insecticide, fungicide, etc.

pH scale scale from 0–14 to measure acidity (potential of hydrogen); it is an inverse log scale such that a lower pH is more acidic and a decrease in pH of 1 point on the scale represents a 10-times increase in acidity

pheromone a chemical communicator which is released by one animal to affect the behaviour of other individuals of the same species

physiology the functioning of a living organism and its parts

phytoplankton small, floating plant-like photosynthetic organisms in aquatic ecosystems

pioneer species the first species to colonize an area after there has been a disturbance

pollinator insects, bats and hummingbirds that distribute pollen between plants

pond a large depression where water collects with its own unique community of plants and animals

pool a small body of still water

population all organisms of a species living in a given area

precipitation all forms of water falling to the ground, such as rain and snow

predator an organism feeding on another living organism (its prey) and killing it

prey switching a predator changing its preference for one prey over another because of changes in prey abundance

primary productivity the amount of energy fixed by plants in the biomass

primary succession succession on a site not containing soil, such as a rock (*see also* succession)

profundal zone deep waters below the limnetic zone

putrification the process of decay or rot

quality of life the condition of having a healthy lifestyle focused on wellness, fitness and spiritual fulfillment

raptor bird of prey, including hawks, owls and eagles

ravine a narrow and deep gorge, typically with a stream at the bottom

retention pond a pond intended to keep water on a site

riffle small rapids

right-of-way a railway corridor for trains, or a utility corridor containing high voltage power lines, or high pressure gas lines that typically are left as green spaces

riparian pertaining to the edge of a river or stream

root-to-shoot ratio the proportion of the weight of roots relative to the weight of the shoots of a plant

saline salty

salmonid belonging to the fish family *Salmonidae,* including salmon and trout

sand small eroded mineral material 0.02–0.2 cm in diameter

savanna a large grassland of coarse grasses with scattered tree growth

scree slope of small stones that have slid down a slope when walked upon

secondary succession succession occurring on a site that already has soil

setback a minimum distance to stay away from a stream or otherwise sensitive area to prevent disturbance

shade-tolerant tree tree species that does well in shade and poorly in open sunlight, typically occurring in a late stage of succession

silt small eroded mineral material 0.002–0.02 cm in diameter

Simpson's index a biodiversity index that considers numbers of species and their overall abundance

Smart Growth a program to contain urban sprawl and encourage principles of sustainability promoted by the Sustainable Communities Network, coordinated by the US Environmental Protection Agency

Snag a standing, dead tree at least 2 m tall

solid waste garbage including paper, wood, organic material, glass and metal

spatial niche the actual area inhabited by a species

specialist having narrow requirements for habitat, food, shelter and breeding sites

species area curve a graph with the numbers of species represented on the Y-axis and area of land on the X-axis

species richness the number of species in an area

standard of living a measure of the amount of material goods available to a person

stewardship looking after a place or something that does not belong to you

street tree a tree on municipal property lining a street

succession refers to the sequential replacement of one community by another in an area, ending with a stable climax community (*see also* climax community)

sun-tolerant tree tree species that does well in open sunlight and poorly in shade, typically occurring in an early stage of succession

sustainable a situation that balances of environmental, social and economic concerns

swamp a wetland with trees and water on or near the surface

symbiosis organisms living together with either positive or negative relationships

taiga, also called coniferous forest or boreal forest a major association of coniferous trees (e.g., spruce, hemlock, cedar), found in the higher latitudes around the globe and characterized by long cold winters and heavy snowfalls

talus sloping mass of small stones or rock fragments

terrestrial pertaining to land

thicket dense shrub area that cannot be seen through

transpiration water lost through the leaves of land plants

trophic level a level in a food web representing organisms obtaining energy by the same number of steps

tundra occurring in a circumpolar distribution at the highest latitudes and characterized by long, cold winters with little precipitation, no trees and a permanently frozen layer of ground called permafrost

turbidity a measure of the cloudiness of water that is usually caused by sediment, micro-organisms and pollutants

understorey the plant community growing above ground but below the canopy of trees consisting of shrubs and low-growing trees

urban pertaining to a city or town defined geographically and politically

urban adaptor species commonly found in rural or more natural environments that have learned to exploit additional resources, such as ornamental vegetation found in moderately developed areas

urban avoider species particularly sensitive to human-induced changes to the landscape and which avoid cities

urban exploiter species that actually prefer disturbed habitat caused by urban development and which have densities are higher in urban environments

urban heat island a region of air warmer than that outside the city that has its own distinct flow pattern

urban landscape gradient the increase in species composition and diversity going from the centre of a city to its periphery

urban runoff water originating from human landscapes through stormdrains or surface flows

variance exceptions granted to normal regulations guiding development within the Official Community Plan

vernal appearing in spring

vestpocket park small garden at an intersection or next to a building

watershed area drained by a particular river or stream

wattle a wall of woven branches, usually of willow that will sprout from its buds, used to stabilize a stream bank

wetland an area with water on or near the surface and vegetation uniquely adapted for the wet conditions

wetted channel width width of water present

wetted depth the existing depth of water in a stream or river
woodlot a small, wooded area usually of closely spaced, fast growing, pioneer species, such as alder, but may contain larger trees with a well-formed canopy
xeriscaping landscaping with plants that require little or no watering
zooplankton small, floating, animal-like heterotrophic organisms in aquatic ecosystems

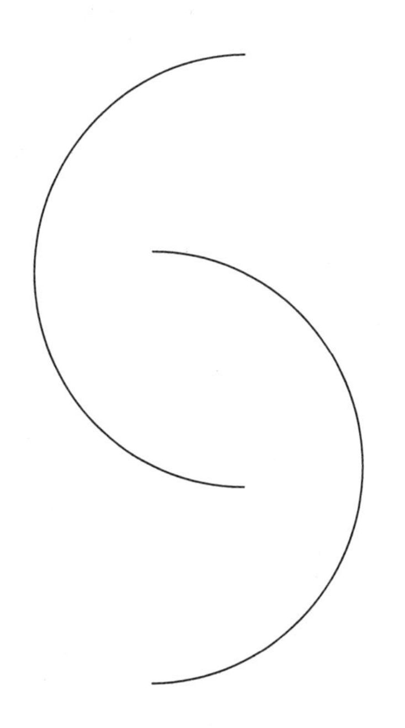

Bibliography

Abbey, B. (1998). *US landscape ordinances: an annotated reference handbook.* New York, NY: John Wiley and Sons, Inc.
Acres International Ltd. (1993). *Boundary Bay area study other environmental Issues component study.* B.C. Ministry of the Environment.
Adams, L.W. & Dove, L.E. (1994). *Wildlife reserves and corridors in the urban environment: a guide to ecological landscape planning and resource conservation.* Maryland, West Virginia: National Institute for Urban Wildlife.
Adams, Mark A, & Whyte. I.A. (1990). *Fish habitat enhancement: a manual for freshwater, estuarine, and marine habitats.* Vancouver, BC: Envirowest Environmental Consultants.
Aguilar, Rosalie. (2003). *BC: special Ppaces project report.* Douglas College Institute of Urban Ecology, 74 pp. plus Appendices.
Ahlbom, A., et al. (1987). *Biological effects of power line fields: New York State power lines project scientific advisory panel final report.* New York, NY: New York State Power Lines Project.
Akbari, H., et al. (Eds.). (1992). *Cooling our communities: a guidebook on tree planting and light-coloured surfacing.* Washington, D.C.: US Environmental Protection Agency.
_____., Rosenfeld, A.H., & H. Taha. (1989). Cooling Urban Heat Islands. In Rodbell, P. (Ed.), *Proceedings of the fourth Urban Forestry Conference.* St. Louis, Missouri: October 15–19, pp 50–57.
Andren, H. (1994). Effect of habitat fragmentation on birds and mammals in landscapes with different proportions of suitable habitat: a review. *Oikos* 71: 355–366.
Andrews, W.A. (Ed). (1973). *Soil ecology.* Scarborough, ON: Prentice Hall.
Argyle, Graham W. and Associates. (1991). Golf Course Development in the Lower Mainland: A Study to Evaluate the Need and Establish Location Criteria. *GVRD Development Services Report*, Burnaby, BC.
Arnold, C. & Gibbons, J. (1996). Impervious surface coverage: the emergence of a key environmental indicator. *J. Amer. Planning Assoc.* 62(2), 243–258.
Bannerman, S. (1998). *Biodiversity and interior habitats: the need to minimize edge effects part 6 of 7.* BC Ministry of Forests Research Program. pp. 1–8.
Barclay, James A. (1992). *Golf in canada: a history*. Toronto, ON: McClelland and Stewart, Inc.
Beckett, P., Freer-Smith, P. & G. Taylor. (2000). Effective tree species for local air quality management. *Journal of Arboriculture*, 26 (1): 12–19.
Bennett, A.F., Henein, K. & G. Merriam. (1994). Corridor use and the elements of corridor quality: chipmunks and fencerows in a farmland mosaic. *Biological Conservation*, 68: 155–165.
Blair, R. (1996). Land Use and Avian species Diversity Along an Urban Gradient. *Ecological Applications,* 6(2), 506–519.
Bocking, Stephen. (2000). Encountering biodiversity: Ecology, Ideas, Action. In: Bocking, Stephen (ed.). *Biodiversity in Canada: Ecology, ideas, action.* Peterborough, ON: Broadview Press, 426 pp.
Bolund, P. & Hunhammar, S. (1999). Ecosystem services in urban areas. *Ecological Economics*, 29: 293–301.

Bradfield, G.E. & Porter, G.L. (1982). Vegetation structure and diversity components of a Fraser estuary tidal marsh. *Canadian Journal of Botany*, 60: 440–451.

Bradley, Gordon A. (1995). Urban Forest Landscapes: Integrating Multidisciplinary Perspectives. In: *Urban forest landscapes: integrating multidisciplinary perspectives.* Univ. Seattle, WA: Washington Press, pp. 3–11.

Bradshaw, A.D. & Chadwick, M.J. (1980). *The restoration of land. studies in ecology, Vol 6.* Oxford and Edinburgh: Blackwell Scientific Publications.

Brewer, Richard. (1994). *The science of ecology*. Saunders Toronto, ON: College Publishing. 773 pp.

BCAgrologist. (1972). Land Use — Are You Concerned? *Delta.* 1(2): 1–4

BC Ministry of Water, Land and Air Protection. (2002). *Environmental Trends in British Columbia.* Victoria, BC: 64 pp.

BC Ministry of Environment, Lands and Parks. (1993). *State of the environment report for British Columbia.* Victoria, BC.

Bright, Alan D., Barro, Susan C. & Randall T. Burtz. (2002). Public attitudes toward ecological restoration in the Chicago metropolitan region. *Soc. And Nat. Res.* 15:763–785.

Brown, Lester R. & Jacobson, Jodi. (1986). The Future of Urbanization: Facing the Ecological and Economic Constraints. *Worldwatch Paper #77.* Washington, D.C.: Worldwatch Institute.

Bullock, Peter, & Gregory, Peter J. (1991). *Soils in the urban environment.* Cambridge, UK: Cambridge University Press.

Burel, F. & Baudry, J. (1995). Social, aesthetic and ecological aspects of hedgerows in rural landscapes as a framework for greenways. *Landscape and Urban Planning.* 33: 327–340.

Butler, R.W. (1999). Winter Abundance and Distribution of Shorebirds and Songbirds on Farmlands on the Fraser River Delta, British Columbia, 1989–1991. *Canadian Field-Naturalist.* 113 (3): 390–395.

_____. (1997). *The great blue heron: a natural history and ecology of a seashore sentinel.* Vancouver, BC.: UBC Press.

_____. (Ed.). (1992). Abundance, Distribution and Conservation of Birds in the Vicinity of Boundary Bay, *BC. Ministry of Environment, Lands and Parks. Technical Report Series* No. 155.

_____. & Campbell, R. Wayne. (1987). The birds of the Fraser River delta: populations, ecology and international significance. *Occasional Paper Number 65,* Canadian Wildlife Service. 73 pp.

Cacik. R. & Schaefer V. (1998). *City of surprises: discovering urban wildlife habitats.* Douglas College Centre for Environmental Studies and Urban Ecology. New Westmintser, BC. 122 pp.

Campbell, S. & Pincott, S. (1995). *Naturescape British Columbia: native plant and animal booklet, Georgia basin*, 60 pages. Victoria: Naturescape British Columbia.

Carr, A. (1999). *Creating bird habitat in small places: a guide for balcony, rooftop, window box and wall plantings using native plants to attract birds.* New Westminster, BC: Douglas College Institute of Urban Ecology, 21 pp.

Carr, L.W. & Fahrig, L. (2001). Effect of Road Traffic on Two Amphibian Species of Differing Vagility. *Conservation Biology.* 15(4): 1071–1078.

Chestnut, Lauraine G. & Rowe, Robert D. (1989). Economic Measures of the Impacts of Air Pollution on Health and Visibility. In *Air pollutant's toll on forests and crops.* James J. MacKenzie and Mohamed T. El-Ashry (Eds.). New Haven, CT: Yale University Press, pp. 316–342.

Chicago Region Biodiversity Council. (2000). *Protecting nature in your community.* Chicago, IL: Northeastern Illinois Planning Commission, 93 pp.

Chicago Region Biodiversity Council. (1999). *Biodiversity recovery plan.* Chicago, IL: Chicago Region Biodiversity Council, 195 pp.

Condon, Patrick (Ed.). (1996). *Sustainable urban landscapes: the surrey design charette.* Vancouer, BC: University of British Columbia, The James Taylor Chair in Landscape & Liveable Environments, 96 pp.

Connell, J.H. (1978). Diversity in tropical rainforests and coral reefs. *Science* 199: 1302–10.

Cowley, J. (1992). Conservationist and developer cooperate to move meadow in Broadmead. *Victoria Naturalist*, 48:4.

Crum, Howard Alvin. (1988). *A focus on peatlands and peat mosses.* Ann Arbor, Mich.: The University of Michigan Press.

Daily, Gretchen. (Ed.). (1997). *Nature's services: societal dependence on natural eosystems*. Washington, DC: Island Press.

Dean, P.B. (1976). Wildlife needs and concerns in urban areas. In: Wiken, E.B. and G.R. Ironside. *Ecological (biophysical) land classification in urban areas.* Fisheries and Environment, Canada.

Devall, Bill & Sessions, George. (1985). *Deep ecology: living as if nature mattered.* Salt Lake City, UT: Peregrine Smith Books, 267 pp.

Dorney, R.S. (1979). The Ecology And Management Of Disturbed Urban Land. *Landscape Architecture*, 69, 268–272, 320.

Duffey, E. et al. (1974). *Grassland ecology and wildlife management.* London, UK: Chapman and Hall.

Dupuis, L. & Waterhouse, F. (2001). Response of Amphibians to Partial Cutting in a Coastal Mixed-Conifer Forest: Management Practices for Retaining Amphibian Habitat in the Vancouver Forest Region. BC Forest Research, Vancouver Forest Region, EN-005, *Wildlife,* March 2001. pp. 1–12.

Dwyer, D. (1991) Economic benefits and costs of urban forests. In Robdell, P.D. (Ed.), *Proceedings of the fifth annual urban forest conference.* Los Angeles, CA: Nov. 15–19, pp. 55–58.

_____., et al. (1992). Assessing the benefits and costs of the urban forest. *Journal of Arboriculture* 18(5): 227–234.

Egler, F.E. (1953). Vegetation management for rights-of-way and roadsides. *Smithsonian Inst. Ann. Report 1953*, pp 299–322.

Elfring, C. (1989). Preserving land through local land trusts. *Bioscience* 39:71–74.

Emmons, L.H. (1984). Geographic variation in densities and diversities of nonflying mammals in Amazonia. *Biotropica* 16: 210–22.

Environment Canada. (1991). *The state of Canada's environment.* Ottawa, ON: Ministry of Supply and Services.

Erickson, Donna. (2004). The relationship of historic city form and contemporary greenway implementationL a comparison of Milwaukee, Wisconsin (USA) and Ottawa, Ontario (Canada). *Landscape and Urban Planning*, 68: 199–221.

Forman, Richard T.T. & Alexander, Lauren E. (1998). Roads and their major ecological effects. *An. Rev. Ecol. and Systematics,* 29: 207–231.

_____. & Gordon M. (1986). *Landscape ecology.* New York, NY: John Wiley and Sons.

_____. & Baudry J. (1984). Hedgerows and hedgerow networks in landscape ecology. *Environmental Management.* 8: 495–510.

Fritz, R., & Merriam, G. (1993). Fencerow habitats for plants moving between farmland forests. *Biological Conservation.* 64: 141–148.

Garber, Steven D. (1987). *The urban naturalist.* New York, NY: John Wiley & Sons, 242 pp.

Georgia Basin Ecosystem Initiative. (2002). *Georgia Basin-Puget Sound ecosystem indicators report.* Vancouver, BC: Environment Canada, Pacific and Yukon Region, 22 pp.

Gibbs, J.P. (2000). Wetland Loss and Biodiversity Conservation. *Conservation Biology.* 14(1): 314–317

Gilbert, O.L. (1991). *The ecology of urban habitats.* London, UK: Chapman & Hall.

Girardet, Herbert. (1992). *The Gaia atlas of cities: new directions for sustainable urban living.* Toronto, ON: Anchor Books.

Glendenning, R. (1959). Biology and control of the coast mole, Scapanus orarius True. *British Columbia. Can. J. Anim. Sci.*, 39: 34–44.

Gobster, P. (1991). Social benefits and costs of enhancing biodiversity in urban forests. In Robdell, P.D. (Ed.), *Proceedings of the Fifth Annual Urban Forest Conference.* Los Angeles, CA: Nov. 15–19, pp. 62–65.

Goldstein, Eric A. & Izeman, Mark A. (1990). *The New York environment book.* Washington, DC: Island Press, 268 pp.

Good, David. (1990). A Green Renaissance. In: Gordon, David. (Ed.). *Green cities: ecologically sound approaches to urban space.* Montreal, PQ: Black Rose Books.

Government of Canada, Fisheries and Oceans. (1980). *Stream enhancement guide.* Vancouver, BC.

Greater Vancouver Regional District. (2003a). Biodiversity Conservation Strategy for the Greater Vancouver Region. Spatial Analysis and Habitat and Species Project Update Workshop. January 21, 2003. Presentation by Susan Haid, Policy and Planning Department, GVRD, Burnaby, BC.

_____. (2003b). Biodiversity Conservation in the Greater Vancouver Region: Issues and Strategic Direction Research(prepared by AXYS Environmental Consulting Ltd). Phase One: Review of Key Biodiversity Conservation Issues, Roles and Responsibilities. Phase Two: Socio-Economic Values of Biodiversity in the Greater Vancouver Region. Phase Three: Priorities, Opportunities and Strategic Directions. Burnaby, BC

_____. (2003). Sustainable Region Initiative: Agriculture and Habitat Task Group. *Discussion Paper.* Burnaby, BC: 10 pp.

_____. (1991). Solid Waste Management Plan Review. Stage 1 Report. Burnaby, BC.

Green, R.E., Osborne, P.E. & E.J. Sears. (1994). The distribution of passerine birds in hedgerows during the breeding season in relation to characteristics of the hedgerow and adjacent farmland. *Journal of Applied Ecology.* 31: 677–692.

Grimm, Nancy B., et al. (2000). Integrated approaches to long term studies of urban ecological systems. *BioSci.* 50(7): 571–584.

Hallam, R. (1973). Memo to K. Wile re: Burrard Seafood Products, dredge sample Sept. 12, 1973. Dept. of Environment, Environmental Protection Service. Unpublished data. Extracted from The Fraser River Estuary Status of Environmental Knowledge to 1974.

Halliwell, R. (2000). Hedgerow management and its impact on the South Devon landscape: evidence from three study areas in the South Hams. Accessed August 16, 2000 from http://www.intellectbooks.com/devon/hertiage/hedgerow/intr.htm.

Hansen, I., Fadg, L. & G. Merriam. (1995). *Mosaic landscapes and ecological processes.* New York, NY: Chapman and Hall, 356 pp.

Hansson, L. (1991). Dispersal and connectivity in metapopulations. *Biological Journal of the Linnean Society*, 42: 89–103.

Hardon, Caroline, et al. (1985). *Belcarra regional park: intertidal and subtidal biophysical inventory.* Vancouver, BC: Douglas College.

Harper, Dave. (1983). Technical Paper Series 5: Trees and Towns. Heritage Conservation Trust.

Harris, L.D. (1988). Edge Effects and Conservation of Biotic Diversity. *Conservation Biology,* 2(4): 330–332.

_____. 1984. *The fragmented forest: island biogeography theory and the preservation of biotic diversity.* Chicago IL: University of Chicago Press, 211 pp.

Harris, R.D. & Taylor E.W. (1973). Human impact on estuarine habitat. Canadian Wildlife Service. Mimeo. *Revision.* 16 pp.

Harrison, Colin. (1984). *A field guide to the nests, eggs and nestlings of North American birds.* Toronto, ON: Collins.

Heath, J. (1981). *Threatened Rhopalocera (Butterflies) in Europe.* Strasbourg, France: Council of Europe.

Healey, M.C. (1997). *Prospects for sustainability: integrative approaches to sustaining the ecosystem function of the Lower Fraser Basin.* Vancouver, BC: Westwater Research Centre and the Sustainable Development Research Institute of the University of British Columbia.

Hecnar S.J. & MCloskey, R.T. (1997). The Effects of Predatory Fish on Amphibian Species Richness and Distribution. *Biological Conservation*, 79:123–131

Heimlich, R.E. & Anderson, W.D. (1987). Dynamics of land use change in urbanizing areas: Experience in the Economic Research Service. In *Sustaining agriculture near cities.* Ankeny, IO: Soil and Water Conservation Society,

Heisler, G. (1986). Energy savings with trees. *Journal of Arboriculture*, 12(5): 113–125.
Hermy, M. & Cornelis, J. (2000). Towards a monitoring method and a number of multifaceted and hierarchical biodiversity indicators for urban and suburban parks. *Landscape and Urban Planning.* 49: 149–162.
Hobbs, R.J. & Saunders, D.A. (1990). Nature conservation: the role of corridors. *Ambio*, 19(2): 94–95.
Hong, San-Kee, et al. (2003). Landscape pattern and its effect on ecosystem functions in Seoul metropolitan area: Urban ecology on distribution of naturalized plant species. *J.Env.Sci.* 15(2): 199–204.
Hoos, Lindsay M. & Packman, Glen A. (1971). The Fraser River Estuary. Status of Environmental Knowledge to 1974. *Report of the Estuary Working Group, Department of the Environment, Regional Board Pacific Region.* Special Estuary Series No. 1. 518 pp.
Hough, Michael. (1984). *City form and natural processes: towards a new urban vernacular.* London. Croom Helm, 281 pp.
Hough, Stansbury and Associates. (1979). *Practical Rooftop Experiment.*
Hull, B.R. & Ulrich R.S. (1991) Health benefits and costs of urban trees. In Robdell, P.D. (Ed.), *Proceedings of the fifth annual urban forest conference.* Los Angeles, Nov. 15–19, 1991, pp. 69–72.
Hutchinson, I. (1982). Vegetation-environment relations in a brackish marsh, Lulu Island, Richmond, B.C. *Canadian Journal of Botany*, 60: 452–462.
Irish Hedgerows. Retreived August 16, 2000 from http://www.iwt.ie/hedges.html.
James, Sandy. (2003). Blooming Boulevards: A New Tool for Green Space Management Context. In: Scott-Ashe, et al. *Changing Places Conference: Challenges in green space management.* New Westminster, BC: Douglas College Institute of Urban Ecology.
Janzen, D.H. (1976). Why are there so many species of insects? *Proc.XV Int. Cong. Ent:* 84–94.
Johnson, A.D. & Gerhold, H.D. (2001). Carbon storage by utility-compatible trees. *Journal of Arboriculture*, 27(2): 57–67.
Johnson, Lorraine. (2002). *Case studies of urban greening partnerships.* Toronto, ON: Evergreen.
Kaplan, Rachel & Kaplan, Stephen. (1995). *The Experience of Nature: A Psychological Perspective.* Ann Arbor, MI: Ulrich's Bookstore, 340 pp.
Keil, Roger & Desfor, Gene. (2003). *Local Environment*, 8(1): 27–44.
Kennedy, Christina, Wilkinson, Jessica & Jennifer Balch. (2003). *Conservation thresholds for land use planners.* Washington, DC: Environmental Law Institute, 55 pp.
King, Cuchlaine A.M. (1972). *Beaches and coasts.* London, UK: Edward Arnold.
Klinka, K., et al. (1989). *Indicator plants of coastal British Columbia.* Vancouver, BC: University of British Columbia Press.
Knopf, Jim, et al. (1995). *Natural gardening.* San Francisco, CA: The Nature Company and Time Life Books, 288 pp.
Kruckberg, Arthur R. (1982). *Gardening with native plants of the Pacific northwest.* Vancouver, BC:

Krajina, V.J. (1969). Biogeoclimatic zones and classification of British Columbia. Ecology of Western North America. *Department of Botany, University of British Columbia, Vancouver*, 2(1): 148–168.

Kuo, Frances & Sullivan, W. (2001). Environment and crime in the inner city: Does vegetation reduce crime? *Environment and Behaviour*, 33: 343–367.

Lanzara, Paola & Pizzetti, Mariella. (1977). *Simon and Shuster's guide to trees.* Toronto, ON: Simon and Shuster.

Laurie, I.C. (Ed.). (1979). *Nature in cities.* Toronto, ON: John Wiley and Sons.

Lee, Jack M., et al. (1989). *Electrical and biological effects of transmission lines: review.* Portland: U.S. Department of Energy Bonneville Power Administration.

Lenanton, John. (1980). *The home gardener.* California: Coast Community Colleges.

Lenat, D.R. & Crawford, J.K. (1994). Effects of land use on water quality and fauna on three North Carolina streams. *Hydrobiologia*, 294: 185–199.

Levy, Johnathan I., et al. (2000). Particle concentrations in urban microenvironments. *Envtl. Health Perspect*, 108(11): 1051–1057.

Levy, Maya, Schaefer, Val. & Kelly Fujibayashi. (1999). *Native plant propagation program techniques manual.* New Westminster, BC: Douglas College Centre for Environmental Studies and Urban Ecology, 109 pp.

Linehan, J., Gross, A. & J. Finn. (1995). Greenway planning: developing a landscape ecological network approach. *Landscape and Urban Planning*, 33: 179–193

Link, Russel. (1999). *Landscaping for wildlife in the Pacific Northwest.* Seattle, WA: University of Washington Press.

Lehmkuhl, J. & Ruggiero, L. (1991). Forest fragmentation in the Pacific Northwest and its potential effects on wildlife. in L. Ruggiero, et al. Wildlife and vegetation of unmanaged Douglas-fir forests. *General Technical Report PNW-GTR-285.* Portland, OR: US Forest Service, pp. 35–46.

Lindroth, C.H. 1957. *The faunal connections between Europe and North America.* New York, NY: John Wiley and Sons, 344 pp.

MacArthur, R.H., & MacArthur, J. (1961). On bird species diversity. *Ecology* 42: 594–98.

MacDonald, D.W. & Johnson, P.J. (1995). The relationship between bird distribution and the botanical and structural characteristics of hedges. *Journal of Applied Ecology*, 32: 492–505.

Maco, S.E. & McPhersonm, E.G. (2002). Assessing canopy cover over streets and sidewalks in street tree populations. *J. Aboriculture*, 28(6): 270–276.

McKinney, Michael L. (2002). Urbanization, Biodiversity and Conservation. *BioSci.* 52(10): 883–890.

McPherson, E. G., et al. (2002). Western Washington and Oregon Community Tree Guide: Benefits, Coast and Strategic Planting. International Society of Arboriculture, Pacific Northwest Chapter, Silverton, Oregon. 76 pp.

_____., Simpson, J.R., and M. Livingston. (1989). Effects of three landscape treatments on residential energy and water use in Tucson, Arizona. *Energy and Buildings*, 13: 127–138.

_____. (1989). Creating an ecological landscape. In Rodbell, P. (ed.). *Proceedings of the fourth urban forestry conference*. St. Louis, MO: October 15–19, pp 63–67.
_____. (1992). Accounting for benefits and costs of urban greenspace. *Landscape and Urban Planning*, 22: 41–51.
_____. & Rowntree, R.A. (1993). Energy conservation potential of urban tree planting. *Journal of Arboriculture*, 19(6): 321–331.
_____. (1994a). Using Urban Forests for energy efficiency and carbon storage. *Journal of Forestry*, 92(10): 36–41.
_____. (1994b). Energy-Saving Potential of Tree in Chicago. In McPherson, E.G., Nowak, D.J., & R.A. Rowntree (Eds). *Chicago's urban forest ecosystem: results of the Chicago urban forest climate project.* Radnor, PA: USDA For. Serv. Northeast For. Exp. Sta, pp. 95–113.
_____. (1998). Atmospheric carbon dioxide reduction by Sacramento's urban forest. *Journal of Arboriculture*, 24(4): 215–223.
_____., J.R. Simpson, Peper, P.J., and Q. Xiao. (1999). Benefit-cost analysis of Modesto's municipal urban forest. *Journal of Arboriculture*, 25(5): 235–248.
Meier, Alan K. (1990/1991). Strategic landscaping and air-conditioning savings: a literature review. *Energy and Buildings*, 15–16: 479–486.
Merilees, B. (1989). *Attracting backyard qildlife: a guide for naturel-overs*. Minnesota: Voyageur Press.
Miller, R.W. (1997). *Urban forestry: planning and managing urban green spaces.* Upper Saddle River, NJ: Prentice Hall.
Miller, M.H., Groenvelt, P.H. & D.P Stonehouse. (1988). Stewardship of Soil and Water in the Food Production System. *Notes on Agriculture*. Guelph ON.: Ontario Agricultural College, pp 5–12.
Ministry of Forests. (1995a). Riparian Management Area Guidebook. Government of BC — Forest Practices Code of BC. Site accessed: Oct. 2002 http://www.for.gov.bc.ca/tasb/legsregs/fpc/fpcguide/riparian/Rip-toc.htm
Moll, G. (1989). The state of our urban forest. American Forest.
Moody, A.I. (1978). *Growth and distribution of the vegetation of a southern Fraser delta marsh.* M.Sc. thesis, University of British Columbia. 153 pp.
Moore, P.D. & Bellamy, D.J. (1974). *Peatlands.* New York, NY: Verlad-Springer.
Morris, David. (1990). The ecological city as a self-reliant city. In Gordon, David. (Ed.). *Green cities: ecologically sound approaches to urban space.* Montreal, PG: Black Rose Books.
Morrison, M.L., Marcot, B.G. & R.W. Mannan. (1992). *Wildlife habitat relationships: concepts and applications.* Seattle, WA: University of Washington Press.
Mosquin, Ted. (2000). Status of and trends in Canadian biodiversity. In Bocking, Stephen (Ed.). *Biodiversity in Canada: ecology, ideas, action.* Peterborough, ON: Broadview Press, 426 pp.
Murphy, D. (1988). Challenges to biodiversity in urban areas. In *Biodiversity.* Wilson, E.O. (Ed.). Washington, DC: National Academy Press.
Musteg, Vernon. (1980). *Landscape gardening.* Toronto, ON: Coles Publishing Co. Ltd.

National Wetlands Working Group. (1988). *Wetlands of Canada*. Ottawa, ON: Ecological Land Classification Series, No. 24 Sustainable Development Branch, Environment Canada and Polyscience Publications Inc.

Niering, W.A., & Egler, F.E. (1955). A shrubland community of *Viburnum lentago* stable for twenty-five years. *Ecology,* 36: 356–360.

Norecol Environmental Consultants, Ltd. (1992). *Environmental management for the Westwood Plateau golf courses.* Vancouver, BC.

North, M.E.A., Dunn, M.W. & J.M. Teversham. (1979). *Vegetation of the Southwestern Fraser Lowland, 1858–1880.* Minister of Supply and Services. Map.

Noss, R. (1987). Protecting natural areas in fragmented landscapes. *Natural Areas Journal*, 7:2–13.

Nowak, D.J. (1991). Urban forest structure and the functions of hydrocarbon emissions and carbon storage. In Robdell, P.D. (Ed.), P*roceedings of the Fifth Annual Urban Forest Conference.* Los Angeles, Nov. 15–19, pp. 48–51.

_____. (1993a). Compensatory value of an urban forest: an application of the tree value formula. *Journal of Arboriculture*, 19(3): 173–177.

_____. (1993b). Atmospheric carbon reductions by urban trees. *Journal of Environmental Management*, 37: 207–217.

_____. & McPherson, E.G. (1993). Quantifying the impacts of trees: the Chicago urban forest climate project. *Unasylva*, 173(44): 39–44.

_____. (1994). Atmospheric carbon dioxide reduction by Chicago's urban forests. In McPherson, E.G., D.J. Nowak, and R.A. Rowntree (Eds). *Chicago's urban forest ecosystem: results of the Chicago urban forest climate project.* Radnor, PA: USDA For. Serv. Northeast For. Exp. Sta., pp. 83–94.

_____. (1994b). Air pollution removal by Chicago's urban forest. pp63–82. In McPherson, E.G., D.J. Nowak, and R.A. Rowntree (Eds). *Chicago's urban forest ecosystem: results of the Chicago urban forest climate project. USDA For. Serv.* Northeast For. Exp. Sta., Radnor, PA

_____. et al. (2000a). A modeling study of the impact of urban trees on ozone. *Atomosheric Environment*, 34: 1601–1613.

_____. (2000b). Tree species selection, design, and management to improve air quality, In *ASLA annual meeting proceedings*, editor D.L. Scheu pp: 23–26

Nowlan, L. & Jeffries, B. (1996). *Protecting British Columbia's wetlands: a citizen's guide.* West coast environmental Law Research Foundation and British Columbia Wetland Network.

O'Connell, M.A. & R.F. Noss. 1992. Private land management for biological conservation. *Environmental Management*, 16(4): 435–450.

Oglesby, Ray T., Carlson, Clarence A., & James A. McCann (Eds.). (1972). *River ecology and man.* New York, NY: Academic Press.

Oke, T. (1976). The significance of the atmosphere in planning human settlements. In Wiken E.B. & G.R. Ironside (Eds.). *Ecological (biophysical) land classification in urban areas.* Fisheries and Environment, Canada.

O'Neil, T.A., et al. (2001). Matrixes for Wildlife-Habitat Relationship in Oregon and Washington. Northwest Habitat Institute. 2001. In D. H. Johnson & T. A. O'Neil (Manag. Dirs.) *Wildlife-habitat relationships in Oregon and Washington.* Corvallis, OR: Oregon State University Press.

Parker, John H. (1983). Landscaping to reduce the energy used in cooling buildings. *Journal of Forestry*, 81: 82–84.
Pearson, Audrey F. (1985). Ecology of Camosun Bog and Recommendation for reforestation. *Technical paper #3.* U.B.C. Technical Committee on the Endowment and Greater Vancouver Regional District Parks Department. 202 pp.
Peck, Sheila. (1998). Planning for biodiversity: issues and examples. Washington, DC: Island Press, 221 pp.
Peterkin, G. F. (1981). *Woodland conservation and Management.* New York, NY: Chapman and Hall.
Pettinger, A. (1996). *Native plants in the coastal garden: a guide for gardeners in British Columbia and the Pacific Northwest.* Vancouver, BC: Whitecap Books.
Pickett, S.T.A, et al. (2001). Urban Ecological Systems: Linking Terrestrial Ecological, Physical and Socioeconomic Components of Metropolitan Areas. *Annu. Rev. Ecol. Syst.*, 32: 127–57.
Pielou, E.C. (1973). *Ecological diversity.* New York, NY: iley-Interscience.
Pinnell, Nadine., Langill, Dave. & Carmelle Birtch. (2004). *Application of a modified street tree inventory technique to street trees in Coquitlam, BC.* New Westminster, BC: Douglas College Institute of Urban Ecology, 39 pp.
Pojar, J., et al. (1994). *Plants of coastal British Columbia (including Washington, Oregon & Alaska).* Vancouver, BC: Lone Pine Publishing.
Pritchard, D.W. (1967). What is an estuary: physical viewpoint. In: Estuaries. (G.H. Lauff, editor). *Amer. Assoc. Adv. Sci.*, (83): 3–5.
Quayle, M. S. Hamilton. (1999). *Corridors of green and gold: impact of riparian greenways on property values.* Vancouver, BC: Report prepared for the Fraser River Action Plan. Department of Fisheries and Oceans Canada.
Reed, R.A., Barnard, J. Johnson, & W.L. Baker. (1996). Fragmentation of a forested Rocky Mountain landscape, 1950–1993. *Biol. Cons.* 25: 267–277.
Reid, George K. (1967). *Pond Life.* New York, NY: Golden Press.
Ricklefs, Robert E. (1990). *Ecology.* New York, NY: W.H. Freeman and Co.
Roseland, Mark. (1997). *Eco-city dimensions: healthy communities, healthy planet.* Gabriola Island, BC: New Society Publishers.
Rubin, E.S., et al. (2002). Bighorn sheep habitat use and selection near an urban environment. *Biological Conservation.* 104(2002): 251–263.
Rudd, H. & Lee, N. (2003). Conserving Biodiversity in the Greater Vancouver Region: Indicator Species and Habitat Quality. Douglas College Institute of Urban Ecology. New Westminster, BC. 84 pp plus Appendices.
_____., Vala, J., & V.H. Schaefer. (2002). The Importance of Backyard Habitat in a Comprehensive Biodiversity Conservation Strategy B A Connectivity Analysis of Urban Green Spaces. *Restoration Ecology*, 10(2): 368–375.
Sandborn, Calvin. (1997). New tools for the green zone, a summary of green sapace and growth: protecting natural areas in BC communities. In *The green zone priorities for tomorrow conference workbook.* Burnaby, BC: Greater Vancouver Regional District, pp. 49–58.
Savage, J. (1987). Greening the City. *Canadian Heritage*, Aug/Sept:31–35.

Schaefer, V.H. (1999a). The Green Links Project: A Holistic Approach to Habitat Restoration in Cities. Ecological Restoration. *Ecological Restoration*, 17(4): 250–251.

_____. (1999b). Green links and urban biodiversity — an experiment in connectivity. London, England: *UrbanNature Magazine*, p. 100–104.

_____. (1994). Urban Biodiversity. In *Biodiversity in British Columbia.* Harding, L. & E. McCullum (Eds.). Environment Canada, pp. 307–318. .

_____. (1984). A theory on the dentition of moles (Talpidae). Discovery 13(2):64–65.

_____. & M. Sulek. (1997). Green Links. Vol. 1 Coquitlam Demonstration Project, 143 pp. Vol. 2 Implementation, Enhancement and Community Action, 320 pp. Vol. 3 Additional Data and Documentation for the Coquitlam Project, 124 pp. Vol. 4 Burnaby Demonstration Project, 176 pp. Vol. 5 Surrey-Delta Demonstration Project, 159 pp. Douglas College Institute of Urban Ecology.

_____. et al. (1992). *Urban Ravines (Vol. 1). Byrne Creek Ravine — A Case Study. (Vol. 2). B.C. Lower Mainland Urban Ravines Inventory.* New Westminster, BC: Douglas College Institute of Urban Ecology.

_____. & J. Grass. (1989). Interpreting Nature: A Primer for Understanding Natural History, Vols. 1 and 2.Douglas College and Greater Vancouver Regional District. 558 pp.

Schippers, P., et al. (1996). Dispersal and habitat connectivity in complex heterogeneous landscapes: an analysis with a GIS-based random walk model. *Ecography*, 19: 97–106.

Scott, K.I., McPherson, E.G. & J.R. Simpson. (1998). Air pollutant uptake by Sacramento's urban forest. *J. Arboriculture*, 24(4): 224–234.

_____., Simpson, J.R. & E.G. McPherson. (1999). Effects of tree cover on parking lot microclimate and vehicle emissions. *J. Aboriculture*, 25(3): 129–141.

Searns, Robert M. (1995). The evolution of greenways as an adaptive urban landscape form. *Landscape and Urban Planning*, 33: 65–80.

Seymour, Murray. (2000). *Toronto's ravines: walking the hidden country.* Erin, ON: The Boston Mills Press, 168 pp.

Smith, Daniel S. & Hellmund, Paul Cawood. (Eds.) (1993). *Ecology of greenways: design and function of linear conservation areas.* Minneapolis, MN: University of Minnesota.

Sears, Paul B. (1969). *Lands beyond the forest.* Englewood Cliffs, NJ: Prentice-Hall.

Simpson, J.R. & McPherson, E.G. (1996). Potential of tree shade for reducing residential energy use in California. *Journal of Arboriculture*, 22(1): 10–18.

_____. (1998). Urban forest impacts on regional cooling and heating energy use: Sacramento County case study. *Journal of Arboriculture*, 24(4): 201–213.

Smith, D.S. (1993). Introduction. In *Ecology of Greenways: Design and Function of Linear Conservation Areas*, D.S. Smith & P.C. Hellmund, (Eds.). Minneapolis, MN: University of Minnesota Press, 222 pp.

Smith, J.M.B., Borgis, S. & V. Seifert. (2000). Australian Geog. Stud. 38(3): 263–274.

Smith, Robert Leo. (1996). *Ecology and field biology, 5e.* New York, NY: Harper and Row.

Sole, Richard., & Goodwin, Brian. (2000). *Signs of life.* New York, NY: Basic Books (Perseus Books Group), 322 pp.

Stewardship Series. (1996). *The streamkeepers handbook: a practical guide to stream and wetland care.* Fisheries and Oceans Canada, Ottawa. 14 Modules.

_____. (1995). *Community greenways: linking communities to country, and people to nature. Written by Lanarc Consultants Ltd.* Victoria, BC: BC Ministry of Environment, Lands and Parks and Ministry of Municipal Affairs, 72 pp.

_____. (1994). *Stream stewardship: a guide for planners and developers. Written by Lanarc Consultants Ltd.* Victoria, BC: BC Ministry of Environment, Lands and Parks and Ministry of Municipal Affairs, 48 pp.

Straley, Gerald B. (1992). *Trees of Vancouver.* Vancouver, BC: UBC Press, 232 pp.

Sunset Books and Sunset Magazine. (1974). *Basic Gardening Illustrated.* Menlo Park, Cal.: Lane Magazine and Book Co.

Taylor, P.D., et al. (1993). Connectivity is a vital element of landscape structure. *Oikos*, 68: 571–573.

Taylor, Terry. (1990). Burns Bog: refuge for ice age plants. *Discovery*, 19(4): 120–121.

Thomas, J.A. (1984). The conservation of butterflies in temperate countries: Paste efforts and lessons for the future. In Vane-Right, R.I. and P.R. Ackery. (Eds.). *The Biology of Butterflies.* London, UK: Academic Press. pp. 333–353.

Ulrich, R.S. (1984). View through a window may influence recovery from surgery. *Science*, 224: 420–421.

U.S. Department of Agriculture. (1977). *Gardening for Food and Fun.* The Yearbook of Agriculture.

van Apeldoorn, R.C., et al. (1992). Effects of habitat fragmentation on the bank vole, Clethrionomys glareolus, in an agricultural landscape. *Oikos*, 65: 265–274.

Vancouver Urban Landscape Task Force. (1992). *Greenways, public ways.* Final report for the City of Vancouver.

Wackernagel, Mathis. & Rees, William. (1996). *Our ecological footprint.* Gabriola Island, BC: New Society Publishers.

Walls, S., Blaustein, A. & J. Beatty. (1992). Amphibian Biodiversity of the Pacific Northwest with Special Reference to Old-Growth Stands. *The Northwest Environmental Journal,* 8:53–69.

Watson, R.A. (1991). Values and Problems of Urban Trees. In Schaefer, V. (Ed.). *Environmental Issues in Greater Victoria.* New Westminster, BC: Douglas College, pp. 145–150.

Watts, S.B. (Ed.). (1983). *Forestry handbook for British Columbia. 4th ed.* Vancouver, BC: Forestry Undergraduate Society, University of British Columbia, 611 pp.

Welsh, H. & Droege, S. (2000). A Case for Using Plethodontid Salamanders for Monitoring Biodiversity and Ecosystem Integrity of North American Forests. *Conservation Biology,* 15(3): June 2002, pp. 558–569.

Wertheim, Anne (1984). *The Intertidal Wilderness.* San Francisco, CA: Sierra Club Books.

Wetzel, Robert G. (1983). *Limnology.* Philadelphia, PA: Saunders College Publishing.

Whiting, Peter. (2000). Economic aspects of Canadian biodiversity. In Bocking, Stephen. (Ed.). Biodiversity in Canada: ecology, ideas, action. Peterborough, ON: Broadview Press, 426 pp.

Wilcox, B.A. & Murphy, D.D. (1985). Conservation strategy: The effects of fragmentation on extinction. *Am.Nat.* 125(6): 879–887.

Wilson, E.O. (1992). *The diversity of life.* Cambridge, MA: The Belknap Press of Harvard University Press, 424 pp.

Wilson, John D. & Dorcas, Micheal E. (2003). Effects of habitat disturbance on stream salamanders: Implications for buffer zones and watershed management. *Cons. Biol.*, 17(3): 763–771.

With, K.A. & Crist, T.O. (1995). Critical Thresholds in Species' Responses to Landscape Structure. *Ecology*, 76(8): 2446–2459.

Xiao, Q., et al. (1998). Rainfall interception by Sacramento's urban forest. *J.Arboric.* 24(4): 235–244.

Yang, Shao-jin, Dong, Jin-quan. & Bing-ru Cheng. Characteristics of air particulate matter and their sources in urban and rural area of Beijing, China. *J. Environ.Sc.* 12(4): 402–409.

Yates, Steve. (1991). Adopting a stream: a northwest handbook. Seattle, WA: An Adopt-A-Stream Foundation Publication. Distributed by University of Washington Press, 116 pp.

Zukowski, H. (1990). Trees on Trial. *Western Living.* May.

www.americanforests.org. June 15, 2001

www.gvrd.bc.ca June 15, 2001

www.tourismvancouver.com (1998 annual report: Emissions inventory update for lower fraser valley airshed 1999 (from gvrd website)

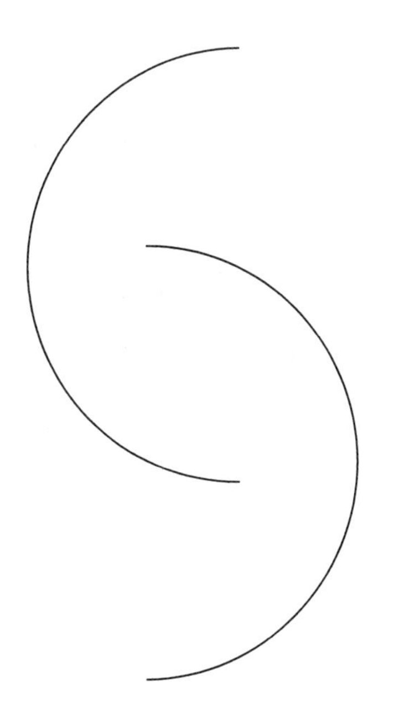

Index

Index

A

B

C

D

E

F

L

M

Q/R

S

T

U